METAHEURISTIC PROCEDURES FOR TRAINING NEURAL NETWORKS

OPERATIONS RESEARCH/COMPUTER SCIENCE INTERFACES SERIES

Other published titles in the series:

Greenberg / *A Computer-Assisted Analysis System for Mathematical Programming Models and Solutions: A User's Guide for ANALYZE*
Greenberg / *Modeling by Object-Driven Linear Elemental Relations: A Users Guide for MODLER*
Brown & Scherer / *Intelligent Scheduling Systems*
Nash & Sofer / *The Impact of Emerging Technologies on Computer Science & Operations Research*
Barth / *Logic-Based 0-1 Constraint Programming*
Jones / *Visualization and Optimization*
Barr, Helgason & Kennington / *Interfaces in Computer Science & Operations Research: Advances in Metaheuristics, Optimization, & Stochastic Modeling Technologies*
Ellacott, Mason & Anderson / *Mathematics of Neural Networks: Models, Algorithms & Applications*
Woodruff / *Advances in Computational & Stochastic Optimization, Logic Programming, and Heuristic Search*
Klein / *Scheduling of Resource-Constrained Projects*
Bierwirth / *Adaptive Search and the Management of Logistics Systems*
Laguna & González-Velarde / *Computing Tools for Modeling, Optimization and Simulation*
Stilman / *Linguistic Geometry: From Search to Construction*
Sakawa / *Genetic Algorithms and Fuzzy Multiobjective Optimization*
Ribeiro & Hansen / *Essays and Surveys in Metaheuristics*
Holsapple, Jacob & Rao / *Business Modelling: Multidisciplinary Approaches — Economics, Operational and Information Systems Perspectives*
Sleezer, Wentling & Cude/*Human Resource Development And Information Technology: Making Global Connections*
Voß & Woodruff / *Optimization Software Class Libraries*
Upadhyaya et al / *Mobile Computing: Implementing Pervasive Information and Communications Technologies*
Reeves & Rowe / *Genetic Algorithms—Principles and Perspectives: A Guide to GA Theory*
Bhargava & Ye / *Computational Modeling And Problem Solving In The Networked World: Interfaces in Computer Science & Operations Research*
Woodruff / *Network Interdiction And Stochastic Integer Programming*
Anandalingam & Raghavan / *Telecommunications Network Design And Management*
Laguna & Martí / *Scatter Search: Methodology And Implementations In C*
Gosavi/ *Simulation-Based Optimization: Parametric Optimization Techniques and Reinforcement Learning*
Koutsoukis & Mitra / *Decision Modelling And Information Systems: The Information Value Chain*
Milano / *Constraint And Integer Programming: Toward a Unified Methodology*
Wilson & Nuzzolo / *Schedule-Based Dynamic Transit Modeling: Theory and Applications*
Golden, Raghavan & Wasil / *The Next Wave in Computing, Optimization, And Decision Technologies*
Rego & Alidaee/ *Metaheuristics Optimization via Memory and Evolution: Tabu Search and Scatter Search*
Kitamura & Kuwahara / *Simulation Approaches in Transportation Analysis: Recent Advances and Challenges*
Ibaraki, Nonobe & Yagiura / *Metaheuristics: Progress as Real Problem Solvers*
Golumbic & Hartman / *Graph Theory, Combinatorics, and Algorithms: Interdisciplinary Applications*
Raghavan & Anandalingam / *Telecommunications Planning: Innovations in Pricing, Network Design and Management*
Mattfeld / *The Management of Transshipment Terminals: Decision Support for Terminal Operations in Finished Vehicle Supply Chains*

METAHEURISTIC PROCEDURES FOR TRAINING NEURAL NETWORKS

Edited by

Enrique Alba and Rafael Martí
University of Málaga; University of Valencia.

Enrique Alba
Universidad de Malaga
Malaga, Spain

Rafael Marti
Univeristat de Valencia
Burjassot, Spain

ISBN-13: 978-1-4419-4128-2

ISBN-10: 0-387-33416-5 (e-book)
ISBN-13: 978-0387-33416-5 (e-book)

Printed on acid-free paper.

9 8 7 6 5 4 3 2 1

springer.com

A toda mi familia

E. Alba

A Jose Javier y a Javier por sus silencios

R. Martí

Contents

Part III: POPULATION BASED METHODS

Part IV: OTHER ADVANCED METHODS

Contributing Authors

Emile Aarts
Philips Research Laboratories, Prof. Holstlaan 4, 5656 AA Eindhoven, The Netherlands.

Enrique Alba
Department of Languages and Computer Science, University of Málaga, Spain.

Ada M. Álvarez
Universidad Autónoma de Nuevo León, Nuevo León, México.

Francisco R. Ángel-Bello
Instituto Tecnológico de Estudios Superiores de Monterrey, Campus Monterrey, Nuevo León, México.

Alberto Aragón
Departamento Economía Aplicada. University of Burgos, Spain.

Christian Blum
ALBCOM, LSI, Universitat Politècnica de Catalunya, Barcelona Spain.

Francisco Chicano
Department of Languages and Computer Science, University of Málaga, Spain.

Bernabé Dorronsoro
Department of Languages and Computer Science, University of Málaga, Spain.

Ignacio J. García del Amo
Grupo de Computación Inteligente, Instituto Universitario de Desarrollo Regional, ETS Ingeniería Informática. Universidad de La Laguna, 38271 La Laguna, Spain.

Nicolás García-Pedrajas
Department of Computing and Numerical Analysis, University of Córdoba, Spain.

Fred Glover
Leeds School of Business, University of Colorado, Campus Box 419, Boulder, CO 80309.

José Luis González-Velarde
Instituto Tecnológico de Estudios Superiores de Monterrey, Campus Monterrey, Nuevo León, México.

César Hervás-Martínez
Department of Computing and Numerical Analysis, University of Córdoba, Spain.

Jan Korst
Philips Research Laboratories, Prof. Holstlaan 4, 5656 AA Eindhoven, The Netherlands

Natalio Krasnogor
School of Computer Science and I.T. University of Nottingham. England.

Manuel Laguna
Leeds School of Business, University of Colorado at Boulder.

Paulo J. G. Lisboa
The Statistics and Neural Computation Research Group, School of Computing and Mathematical Sciences, Liverpool John Moores University, United Kingdom.

Julio Madera
Department of Computing, University of Camagüey, Cuba.

Rafael Martí
Dpto. de Estadística e Investigación Operativa, Universidad de Valencia, Dr. Moliner 50, 46100 Burjassot (Valencia) Spain.

José David Martín
Grupo de Procesado Digital de Señales, Dpt. Enginyeria Electrònica, Escola Tècnica Superior d'Enginyeria, Universitat de València, Spain.

Belén Melián Batista
Grupo de Computación Inteligente, Instituto Universitario de Desarrollo Regional, ETS Ingeniería Informática. Universidad de La Laguna, 38271 La Laguna, Spain.

Wil Michiels
Philips Research Laboratories, Prof. Holstlaan 4, 5656 AA Eindhoven, The Netherlands.

Nenad Mladenović
School of Mathematics, Brunel University, West London, United Kingdom.

José Andrés Moreno Pérez
Grupo de Computación Inteligente, Instituto Universitario de Desarrollo Regional, ETS Ingeniería Informática. Universidad de La Laguna, 38271 La Laguna, Spain.

Domingo Ortiz-Boyer
Department of Computing and Numerical Analysis, University of Córdoba, Spain.

Joaquín Pacheco
Departamento Economía Aplicada. University of Burgos, Spain.

Krzysztof Socha
IRIDIA, Université Libre de Bruxelles, Brussels, Belgium.

Herman Sontrop
Philips Research Laboratories, Prof. Holstlaan 4, 5656 AA Eindhoven, The Netherlands.

Emilio Soria
Grupo de Procesado Digital de Señales, Dpt. Enginyeria Electrònica, Escola Tècnica Superior d'Enginyeria, Universitat de València, Spain.

Pieter van der Horn
Philips Research Laboratories, Prof. Holstlaan 4, 5656 AA Eindhoven, The Netherlands.

Preface

Artificial neural networks (ANNs) offer a general framework for representing non-linear mappings from several input variables to several output variables, and they can be considered as an extension of the many conventional mapping techniques. In addition to many considerations on their biological foundations and their really wide spectrum of applications, constructing appropriate ANNs can be seen as a really hard problem. A distinguished task in building ANNs is the tuning of a set of parameters known as *weights*. This will be the main focus of the present book. The trained ANNs can be later used in classification (or recognition) problems, where the ANN outputs represent categories, or in prediction (approximation) problems, where the outputs represent continuous variables.

In the process of training the ANN (*supervised learning*), the problem is to find the values of the weights that minimize the error across a set of input/output pairs (patterns) called the training set. In a first stage, the training is an unconstrained nonlinear optimization problem, where the decision variables are the weights and the objective is to reduce the training error. However, the main goal in the design and training of ANNs is to obtain a model which makes good predictions for new inputs (which is termed as *generalization*). Therefore the trained ANN must capture the systematic aspects of the training data rather than their specific details. Hence, as it has been well documented, the optimization problem involved in the training/generalization process is of an extreme hardness.

Metaheuristics provide a means for solving complex optimization problems to obtain acceptable solutions or even global optima. These methods are designed to search for such global optima in complex problems where other mechanisms fail because: the problem is ill-defined, or has a very large dimensionality, or a high interaction between variables exists, or require unaffordable computational efforts for exact methods. Experimental testing of metaheuristics show that the search strategies embedded in such procedures are capable of finding solutions of high quality to hard problems in industry, business, and science within reasonable computational time. The tools and mechanisms that have emerged from the creation of metaheuristic methods have also proved to be remarkably efficient, resulting in what has been coined as hybrid methods.

Apart from some sparse efforts to bring together metaheuristic techniques to train ANNs (which include conference sessions on this field), there is no a single source of reference for such goal. In this book we aim at giving a unified approach to the work of training ANNs with modern heuristics, given the overwhelming literature proving their appropriateness to escape local optima and to solve problems in very different mathematical scenarios (two features that encapsulate important shortcomings of other well-known algorithms specifically designed to train ANNs).

The book's goal is to provide successful implementations of metaheuristic methods for neural network training. Moreover, the basic principles and fundamental ideas given in the book will allow the readers to create successful training methods on their own. Apart from Chapter 1, in which classical training methods are reviewed for the sake of the book's completeness, we have classified the chapters in three main categories. The first one is devoted to *local search based* methods, in which we include Simulated Annealing, Tabu Search, and Variable Neighborhood Search. The second part of the book presents the most effective *population based* methods, such as Estimation Distribution algorithms, Scatter Search, and Genetic Algorithms. Finally, the third part includes other advanced techniques, such as Ant Colony Optimization, Co-evolutionary methods, GRASP, and Memetic algorithms. All these methods have been shown to work out high quality solutions in a wide range of hard optimization problems, while in this book we restrict our attention to their application to the ANN training problem.

This book is engineered to provide the reader with a broad coverage of the concepts, methods, and tools of this important area of ANNs within the realm of continuous optimization. In fact, many applications dealing with continuous spaces could profit from the advances described in it. The chapters can be addressed separately depending on the reader's necessities. It would be of interest to researchers and practitioners not only in neural networks but also in management science, economics, and engineering in general. Besides, it can be used as a textbook in a master course, a doctoral seminar, or as a reference book for computer science in areas such as enterprise resource planning and supply chain management.

We hereby acknowledge the help of Springer's editorial office in all this work, as well as all the support from many people in our daily working environment. We thank very especially all the contributors, who made a huge effort to have state-of-the-art solutions in this book applied to interesting academic and industrial applications.

Enrique Alba and Rafael Martí

ACKNOWLEDGEMENTS

The research of Enrique Alba is partially funded by the *Ministerio de Educación y Ciencia* and the Regional Development European Found (FEDER) under contracts TIC2002-04498-C05-02 (the TRACER project, http://tracer.lcc.uma.es) and TIN2005-08818-C04-01 (the OPLINK project).

The research of Rafael Martí is partially supported by the *Ministerio de Educación y Ciencia* (ref. TIC2003-C05-01 and TIN2004-20061-E) and by the *Agencia Valenciana de Ciència i Tecnologia* (ref. GRUPOS03/189).

Chapter 1

CLASSICAL TRAINING METHODS

Emilio Soria[1], José David Martín[1] and Paulo J. G. Lisboa[2]
[1] *Grupo de Procesado Digital de Señales, Dpt. Enginyeria Electrònica, Escola Tècnica Superior d'Enginyeria, Universitat de València, Spain.*
[2] *The Statistics and Neural Computation Research Group, School of Computing and Mathematical Sciences, Liverpool John Moores University, United Kingdom.*

Abstract: This chapter reviews classical training methods for multilayer neural networks. These methods are widely used for classification and function modelling tasks. Nevertheless, they show a number of flaws or drawbacks that should be addressed in the development of such systems. They work by searching the minimum of an error function which defines the optimal behaviour of the neural network. Different standard problems are used to show the capabilities of these models; in particular, we have benchmarked the algorithms in a non-linear classification problem and in three function modelling problems.

Key words: Multilayer perceptron; delta rule; cost function.

1. INTRODUCTION

There are two main approaches to describe Artificial Neural Networks (ANNs). Some authors describe ANNs as biological models that can be applied to engineering problems (Arbib, 2003). However, other authors consider ANNs as mathematical models related to statistical models, either linear or non-linear (Bishop, 1995; Ripley, 1996; Duda, et al., 2001). These two approaches have coexisted in the theory of neural models from the very beginning, so that advances in this theory have come from both approaches. Biological models have provided the inspiration for the development of new artificial neural models while mathematical and statistical frameworks have consolidated their practical value. This chapter is focused on the mathematical approach of artificial neural models. In this approach, an artificial neural system can be described as it is shown in Figure 1-1.

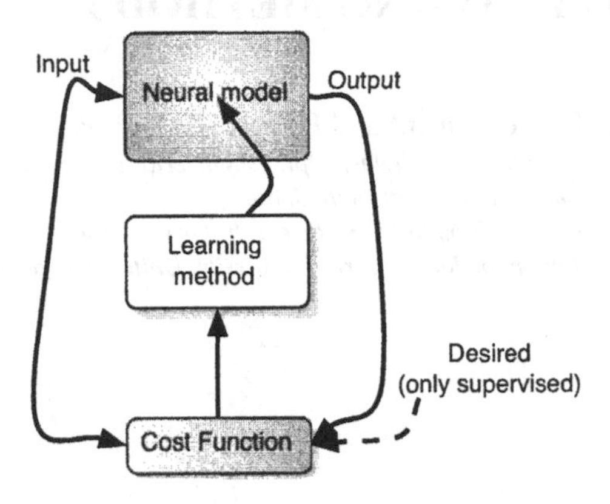

Figure 1-1. Schematic of a neural model.

Figure 1-1 shows a neural model. Three main parts can be observed:

- *Neural model.* It is the structure used to solve a certain problem. This model can be either linear or non-linear, it can have one or more than one outputs, it can consist of a combination (linear, non-linear, hierarchical, etc.) of simple neural models, etc. The performance of this model depends on several parameters which determine the complexity of the neural model.
- *Cost function.* It provides an evaluation about the quality of the solution obtained by the neural model. If an external signal to carry out this evaluation is available, then the neural model is called supervised model. If an external signal is not avaliable, then the neural model is called unsupervised model (Haykin, 1999).
- *Learning method.* The task of the learning algorithm is to obtain those parameters of the neural models that provide a solution to the tackled problem. If this solution does not exist, then the task of the learning algorithm is to find an optimal solution according to the criterion set by the cost function.

Summarizing, there are three basic elements in an artificial neural model. Since all these elements can change independently to the others, there is a huge amount of models/cost functions/ learning algorithms (Arbib, 2003). In this chapter, we focus on the most widely used neural model, the Multilayer Perceptron (MLP) trained by supervised algorithms (in the case of unsupervised training, the resulting model is the so-called *Nonlinear Principal Component Analysis, NPCA*). We also analyse the most common cost functions associated with MLPs. For a further analysis of the most important neural models, supervised and unsupervised, the reader is encouraged to consult the excellent text (Haykin, 1999).

2. MULTILAYER PERCEPTRON

In this Section, the Multilayer Perceptron (MLP) is described. It is probably the most widely used neural network due to its interesting characteristics: universal function approximator and non-linear classifier. The MLP has shown excellent results in many different applications. In Subsection 2.1, the elementary units which form the MLP, the so-called neurons, are presented. Next subsections are devoted to explain how it works; in particular, in Subsection 2.2, the MLP architecture is analysed, in Subsections 2.3 and 2.4., the cost function, i.e., the error function that must be minimised is described, and in Section 2.5., a Bayesian approach to MLP learning is presented. Section 3 analyses the learning algorithms used to carry out the cost function minimisation.

2.1 Neurons

An MLP is a neural network made up by multiple, similar, non-linear processing units, termed neurons by analogy to the integrate-and-fire action of neural cells in biological nervous systems. Each neuron carries out a many-to-one mapping from its inputs to a single scalar output. A general schematic of a neuron model is shown in Figure 1-2 (Hecht-Nielsen, 1989).

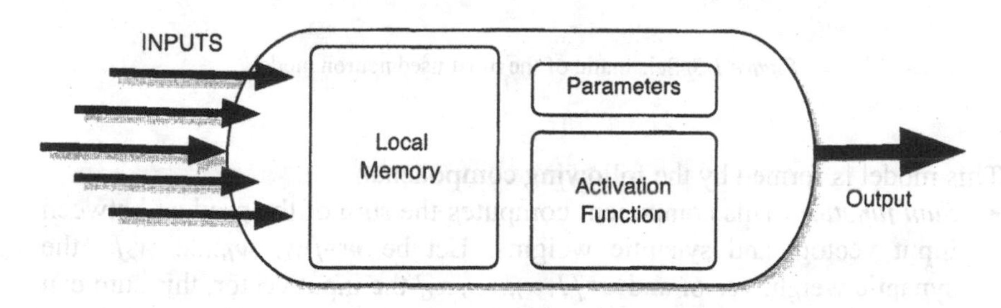

Figure 1-2. General schematic of a neuron.

This neuron model is formed by the following components:

- *Parameters*. These parameters are usually known as *synaptic weights*. They are used in combination with the inputs according to a certain function. Considering input vectors and coefficient vectors, the most used functions are the scalar product and the Euclidean distance.
- *Activation function*. This function is very important since the neural network capabilities to solve complex problems stem from it. It is a non-linear function whose argument is the aforementioned combination between synaptic weights and input vector. The most used functions are the *sign function*, the *hyperbolic tangent*, the *sigmoidal function* (hyperbolic tangent modified by restricting the range of values between 0 and 1), and the *Gaussian function*.
- *Local memory*. This component is used when neural networks are designed to time series modelling. Local memories can store either previous inputs or previous outputs, so that the neuron "remembers" previous behaviours. There are many different possibilities to implement this kind of memories (Weigend & Gershenfeld, 1993).

In the remainder of the chapter, we will focus on the neuron model shown in Figure 1-3:

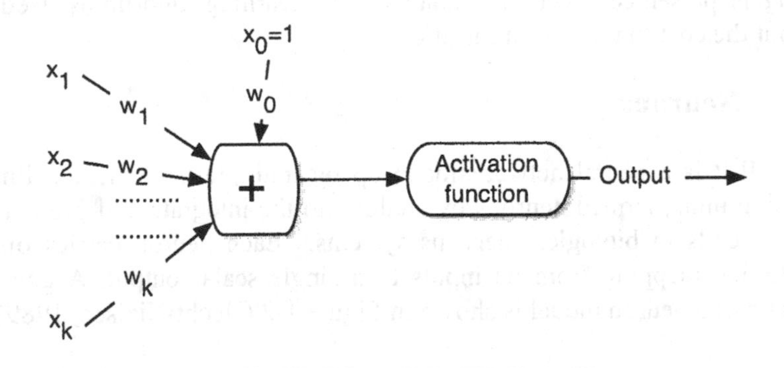

Figure 1-3. Schematic of the most used neuron model.

This model is formed by the following components:

- *Sum function*. This component computes the sum of the product between input vectors and synaptic weights. Let be $\boldsymbol{w}=[w_0, w_1,....., w_k]$ the synaptic weight vector and $\boldsymbol{x}=[1, x_1,....., x_k]$ the input vector, this sum can be given by the scalar product of both vectors. From a biological point of view, this function represents the action of the inputs produced in the axons of the biological neurons (Arbib, 2003). The coefficient w_0 plays a relevant role in the processing of the neuron, and it is called threshold or bias.

- *Activation function.* Among all the possible choices for this activation function, the following ones should be emphasised:

  Sign function. It was the first used activation function in an artificial neural model (Arbib, 2003). This function proposes a crisp separation of the input data, so that they are classified as ±1. This function is defined as follows: $f(x) = \begin{cases} -1 \text{ si } x < 0 \\ 1 \text{ si } x \geq 0 \end{cases}$.

  Sigmoidal function. Since many learning algorithms need a differentiable activation function, and the sign function can not be differentiated in the origin, the sigmoidal function was proposed. This function is given by the following expression: $f(x) = \frac{a}{1+e^{-b \cdot x}}$ where a stands for the amplitude of the function and b is the slope in the origin; the higher the value of b, the closer the sigmoidal function to the sign function (but with the outputs ranging between *0* and a). The bipolar version of the sigmoidal function (outputs ranging between $-a$ and $+a$) is given by the hyperbolic tangent: $f(x) = a \cdot \frac{1-e^{-b \cdot x}}{1+e^{-b \cdot x}}$.

  Gaussian function. An alternative activation function is used in a different type of neural networks, known as *Radial Basis Function (RBF)*. Its expression is given by: $f(x) = K_1 \cdot e^{-K_2 \cdot (x-c)^2}$, being K_1, K_2 and c, values which determine the amplitude of the Gaussian function, its width and its centre, respectively. The main characteristic of this activation function, which makes different from others, is its locality. The function tends to zero from a certain value x on; this way, only in a reduced range of input values, the output of the activation function is considerably different from zero. Gaussian functions have their centre and width values typically heuristically determined from the data.

2.2 Architecture

The architecture of a neural model gives information about how neurons are arranged in the neural model. In the case of MLP, as shown in Figure 1-

4, neurons are arranged in layers: one input layer, one output layer, and one or more than one hidden layers (Arbib, 2003). This arrangement in layers is due to the fact that the outputs of the neurons of a certain layer are used as inputs to the neurons of next layer (network without feedback) and/or to neurons of previous layers (networks with feedback or recurrent networks).

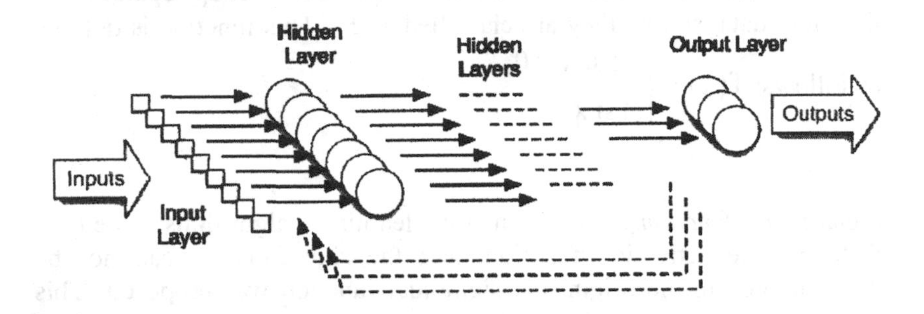

Figure 1-4. Multilayer perceptron. Dotted arrows stand for recurrent neural systems.

The number of neurons in both input and output layers is given by the problem to solve, whereas the number of hidden layers, and the neurons in each layer should be chosen by the designer of the neural network. It is shown that an MLP with one hidden layer is a universal approximator of continuous functions; if the function to model is not continuous, then two hidden layers become necessary (Cybenko, 1988). Although there are several rules to assess the approximately optimal number of hidden neurons, there is no precise methodology to determine this number exactly. Therefore, trial-and-error procedures are usually carried out to estimate the number of hidden neurons, often relying on cross-validation[1] or other evaluation procedures.

2.3 Cost Function

A neural system should be designed to present a desired behaviour, hence, it is necessary to define what is desired, being used a cost function for this task. Two points of view can be considered for cost functions: operational and probabilistic.

- *Operational.* This point of view consists of implementing the final goal of the neural system in the cost function. This is the case of supervised

[1] The cross-validation procedure is described in Section 3.1.

systems; in these systems, the model is designed to provide an output as similar as possible to a certain desired signal (Figure 1-1). An error signal is defined from the difference between the neural model output and the desired values. The goal of the neural model is to obtain an error signal equal to zero; therefore, if an error function is defined provided that its minimum corresponds to a zero error, then the goal is transformed in a problem of function minimisation (Hassoun, 1995; Bishop, 1995).

- *Probabilistic.* Two kind of problems can be solved by using this point of view: function modelling and pattern classification (Haykin, 1999). In the former case, the probability distribution of the desired signals conditioned by the input variables to the neural model must be modelled. In a pattern classification problem, the aim is to model the conditioned probabilities of every class, also by the input variables to the neural model (Bishop, 1995; Ripley, 1996).

The two points of view can be related using the *maximum likelihood principle*, so that assuming different probabilistic models for the error between the network output and the target values, different cost functions are obtained (Bishop, 1995). For instance, if the error is assumed to follow a Gaussian distribution, then the most used cost function is obtained, the mean-square error (L_2 in Table 1-1). The most used cost functions are shown in Table 1-1 (Cichocki & Amari, 2002; Hassoun, 1995; Bishop 1995).

Table 1-1. Most used cost functions. The parameter β controls the error ranges in which a certain cost function is used; this parameter is used in those cases in which different subfunctions are defined for a certain cost function. In the case of the logistic cost function, α controls the robustness to outliers. In the remainder of the chapter, the desired signal will be denoted by d, the output of the neural model by o and the error signal by e, being $e=d-o$

Name	Cost Function
L_2	e^2
L_1	$\lvert e \rvert$
L_p	$\frac{1}{p} \cdot \lvert e \rvert^p$

Name	Cost Function
Entropic	$d \cdot \log\left(\frac{d}{o}\right) + (1-d) \cdot \log\left(\frac{1-d}{1-o}\right)$ $0 \leq o \leq 1$
Logistic	$\frac{1}{\alpha} \cdot \log(\cosh(\alpha \cdot e))$
Huber	$\begin{cases} 0.5 \cdot e^2 & for \lvert e \rvert \leq \beta \\ \beta \cdot \lvert e \rvert - 0.5 \cdot \beta^2 & otherwise \end{cases}$
Talvar	$\begin{cases} 0.5 \cdot e^2 & for \lvert e \rvert \leq \beta \\ 0.5 \cdot \beta^2 & otherwise \end{cases}$

2.4 Relevance of an Adequate Cost Function

The MLP architecture may be configured in two ways, namely for regression, and classification. In the first instance, the hidden layer is nonlinear but the output layer is normally linear and the appropriate cost function is the sum of square errors, reflecting the assumption that the noise in that data is homoscedatic (i.e. uniform across the range of input values) and normally distributed about the origin (Ripley, 1996; Bishop, 1995).

However, if the model is intended for classification, then the output layer becomes non-linear. For binary classification this will be a sigmoid, while for multi-class assignments it will be a softmax, which is a multivariate extension of the sigmoid function. In either case, the appropriate cost function to use is entropic (Ripley, 1996; Bishop, 1995). This can be seen by means of the following example.

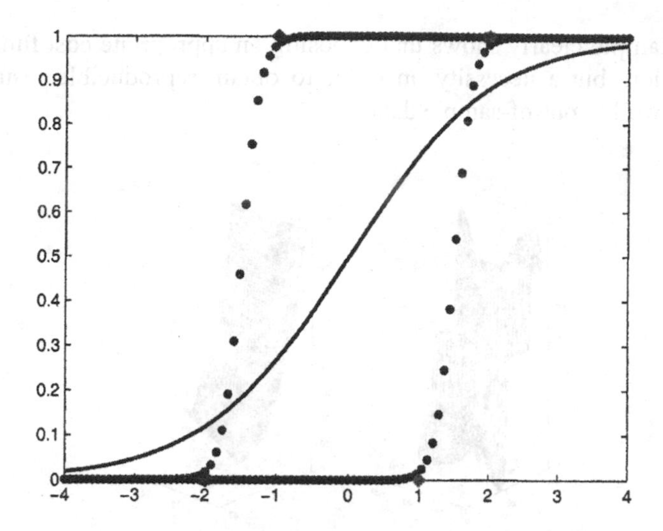

Figure 1-5. The solid line represents the optimal response function for binary classification in the presence of noise, while the dotted lines are two local minima that arise if the incorrect cost-function is used. These local minima may have less cost than the correct fit to the data, leading to inconsistency and inaccurate results.

Consider a model of a single neuron with a single input node directly linked to a single output node. Since, this model has only two parameters, its weight and bias terms, the cost function can be represented for the complete parameter space as a 3D plot. Now, consider the typical problem of classification with noisy data, shown in Figure 1-5. In this case there are 'true' class data, on-class to the right of the origin and off-class to the left, but there are also noisy data that overlap into the wrong side of the decision boundary, as is often the case in practice.

It is now a simple matter to plot the cost function for a sum-of squares error and for a log-likelihood error, as a function of the model parameters. These plots are in Figure 1-6a and 1-6b. It is straightforward to show that as the ratio of true class data to noisy data increases, the local minima along the two ravines spreading diagonally across the plot become closer in value to the correct minimum at the centre of the graph, while retaining a lower cost (i.e. being more optimal) than the true minimum. These ravines correspond to the functions shown as dotted lines in Figure 1-5. Fig. 1-6 (a) also shows the presence of plateaus even in this simple 1-D problem, a situation that can become much more acute in higher dimensions.

This example clearly shows that choosing an appropriate cost function is not an option, but a necessity, in order to obtain reproducible results that generalise well to out-of-sample data.

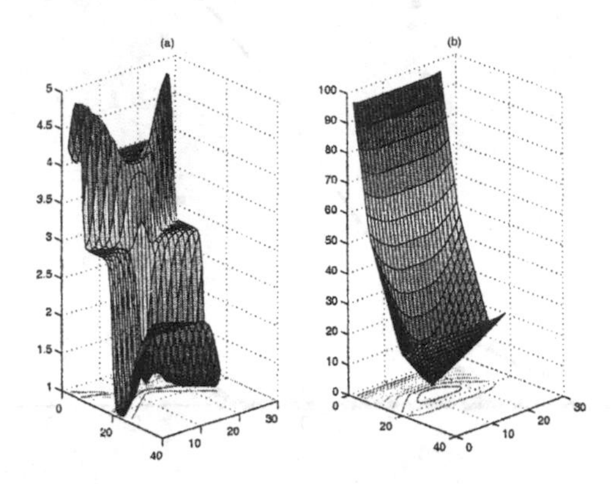

Figure 1-6. Complete maps of the cost function for the data in Fig. 1-5, using a sum of squares error (a) and a log-likelihood error (b). Two quite different surfaces are obtained, one with local minima and the other without.

2.5 Bayesian Approach to MLP Learning

MLP learning can be analysed from a Bayesian pint of view (MacKay, 2003; Bishop 1995), in which the learning process can be seen as an inference process. Since MLP learning is based on mapping an input space into an output space, Bayes' Theorem can be formulated, as follows

$$P(w \mid D) = \frac{P(D \mid w) \cdot P(w)}{P(D)}$$

where D is the desired data set and w is the set of network parameters (synaptic weights). It is necessary to assume a distribution for the neural model parameters; usually, the following expression is used (MacKay, 2003):

$$P(w)=\frac{e^{(-\alpha\cdot E_w)}}{Z_W(\alpha)}$$

E_w is an increasing function of the synaptic-weight values and α is a parameter known as *hyperparameter*. Function $Z_w(\alpha)$ is only used for normalisation purposes. If $P(D|w)$ is a similar distribution to that shown by synaptic weights in Bayes' expression, then:

$$P(D\mid w)=\frac{e^{(-\beta\cdot E_D)}}{Z_D(\beta)}$$

And replacing in the first expression:

$$P(w\mid D)=\frac{e^{(-\alpha\cdot E_W-\beta\cdot E_D)}}{Z_D(\beta)\cdot Z_W(\alpha)}=\frac{e^{(-M(w))}}{Z_M}$$

Taking logarithms and maximising this expression, the conclusion is that the Bayesian objective becomes the minimisation of a cost function, which is given by E_D, together with a regularisation term of the network parameters (E_W). The Bayesian approach has several advantages over approaches based on cost functions (MacKay, 2003; Bishop, 1995).

3. CLASSICAL LEARNING ALGORITHMS

A learning algorithm is a procedure which adapts the parameters and/or architecture of the neural model in order to solve a certain problem. The algorithms to adapt the architecture of the MLP are known as pruning methods or growing methods, depending on the strategy to carry out this adaptation (Haykin, 1999). Pruning methods have been more widely used than growing methods.

Learning algorithms to adapt the parameters of the neural model tend to be based on the minimisation of the cost function chosen to solve the posed problem. First neural models were formed by only one neuron and they used L_2 as cost function, so that minimum was obtained by solving a system of linear equations (Haykin, 1996). However, the practical application of these models involved solving this system of equations in every instant of processing; taking into account the huge number of unknown quantities and the technological state of those years (1950-1960), other approaches were researched in order to solve these systems of equations.

The most important characteristics that a learning algorithm must show, are the following (Haykin, 1999):

- *Efficiency*. Ability to solve the problem with the minimum computacional burden.
- *Robustness*. The algorithm should present immunity to undesired noise.
- *Independence on the initial conditions*. The algorithm should show similar solutions independently of the values used to initialise the algorithm.
- *Generalisation capabilities*. The algorithm should provide the adequate outputs when inputs different to the training data set, are used.
- *Scalability with the size and complexity of the data*. The algorithm should have a computational burden that does not strongly depend on the dimensionality and size of the problem to solve.

3.1 Backpropagation Algorithm (BP)

The BP was the first algorithm used for MLP adaptation (Werbos, 1974; LeCun, 1985; Rumelhart, 1986), and it is still the most known and used nowadays. The goal pursued by the algorithm is to obtain the minimum of the cost function (Section 2.3 of this chapter), which is denoted as *J*. The actual solution of *J* is computationally unfeasible in many applications. Moreover, the solution of *J* in a certain instant may be uninteresting in many cases since the properties of the input signals to the neural model may be time-dependent and the neural models must adapt to these changes. An iterative solution is proposed:

$$w_{n+1} = w_n + \Delta w_n \tag{1}$$

where w are the parameters of the neural model and the subscript stands for the time instant. Many algorithms are designed to accomplish Eq. (1); within these algorithms, there are two possibilities for learning (Haykin, 1999):

- *On-Line*. The MLP is fed during all the training process with the input of every pattern and the corresponding desired output. Then, the error is measured and the synaptic weights are adapted depending on this error by using the chosen algorithm.
- *Batch*. In this case, the error between the network output and the desired values is computed for all the patterns. Then, the model parameters are adapted depending on the average error for all the patterns. The computation of all the outputs of the neural model for all the available input patterns is known as *epoch*.

One of the keys of Eq. (1) is to find out the optimal increase/decrease in the parameters that enables to find the minimum of the function. A low computational burden approach is based on a geometrical analysis of the problem. This analysis is based on finding the direction of the minimum from a certain point w_n in the parameter space of the neural model. Since the function gradient points to the direction of the function maximum, the approach will be based on finding the function minimum by moving the synaptic weights in the opposite gradient direction (Bishop, 1995; Luenberger, 1984):

$$w_{n+1} = w_n - \alpha \cdot \nabla_{w_n} J \tag{2}$$

where α is the so-called learning rate or adaptation constant.

The BP algorithm, based on Eq. (2), is a gradient-descent algorithm which backpropagates the error signals from the output layer to the input layer, thus optimising the values of the synaptic weights through an iterative process. Therefore, two stages can be considered:

- ***Feed-forward propagation***: The output of the neural network is obtained, and then, the error is computed by comparing this output with the desired signal.
- ***Backpropagation***: Depending on the error between the network output and the desired values, the algorithm optimises the values of the synaptic weights by means of error backpropagation from the output layer to the input layer, and through the hidden layers.

A schematic of a general connection between two neurons (i,j) is shown in Figure 1-7.

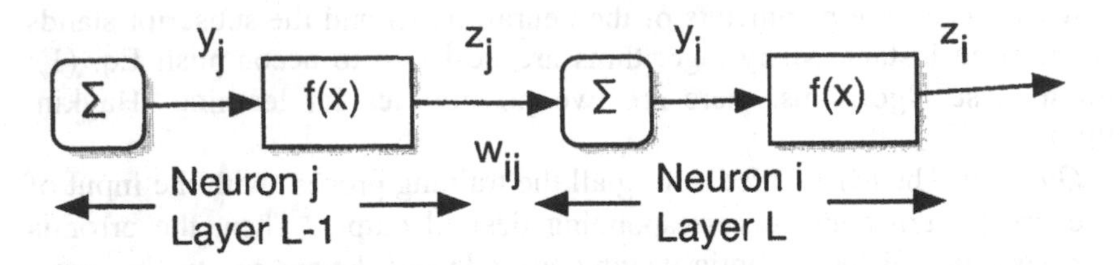

Figure 1-7. Schematic of a connection between two neurons.

The gradient of the cost function must be computed with respect to the parameters of the neural network, i.e., the synaptic weights. The following notation will be used: $y_s^{L-1}; z_s^{L-1}$ stands for the output of the sum function and the activation function, respectively (neuron *s* and layer *L-1*). Therefore, taking all these facts into account:

$$\frac{\partial J}{\partial w_{ij}^L} = \frac{\partial J}{\partial y_i^L} \cdot \frac{\partial y_i^L}{\partial w_{ij}^L} = \delta_i^L \cdot \frac{\partial}{\partial w_{ij}^L} \sum_s w_{is}^L \cdot z_s^{L-1} = \delta_i^L \cdot z_j^{L-1} \qquad (3)$$

$$\text{where } \delta_i^L = \frac{\partial J}{\partial y_i^L}$$

δ is known as local gradient, an can be obtained as follows:

$$\delta_i^L = \frac{\partial J}{\partial y_i^L} = \sum_m \frac{\partial J}{\partial y_m^{L+1}} \cdot \frac{\partial y_m^{L+1}}{\partial y_i^L} = \sum_m \delta_m^{L+1} \cdot \frac{\partial y_m^{L+1}}{\partial y_i^L} \Rightarrow$$

$$\delta_i^L = \sum_m \delta_m^{L+1} \cdot \frac{\partial}{\partial y_i^L} \cdot \sum_j w_{mj}^{L+1} \cdot z_j^L \Rightarrow \delta_i^L = \frac{dz_i^L}{dy_i^L} \cdot \sum_m \delta_m^{L+1} \cdot w_{mi}^{L+1} \qquad (4)$$

Therefore, an iterative procedure is used to compute local gradients of first layers from local gradients of last layers. The output layer gradient is given by:

$$\delta_k = \frac{\partial J}{\partial e_k} \cdot \frac{de_k}{dz_k} \cdot \frac{dz_k}{dy_k} = -\frac{\partial J}{\partial e_k} \cdot \frac{dz_k}{dy_k} \qquad (5)$$

Output gradient depends on the error and activation functions. A very attractive advantage of using a sigmoidal/hyperbolic tangent activation function stems from the fact that their derivatives can be expressed by using the own activation functions (Haykin, 1999).

In spite of the advantages offered by the BP algorithm, it also shows a number of drawbacks that should be known (Haykin, 1999; Arbib, 2003; Bishop 1995; Orr & Müller, 1998):

- *Neuron saturation.* Since the derivative of the activation function appears in the weight update, and this derivative equals zero for the most used activation functions (hyperbolic tangent and sigmoidal) in the function extremes, weights are not updated in these zones although the modelling error is different from zero.
- *Weight initialisation.* Weight initialisation is basic in order to achieve a good modelling. Since the learning algorithm is based on finding the minimum of the error function that is closest to a certain initial point, this minimum may be a local minimum, not a global one, and therefore this initial point is fundamental for ensuring that the network finally achieves a global minimum. Weight initialisation affects algorithm behaviour in three main factors a) *Convergence speed;* the convergence speed depends not only on the learning rate, but also on the initial distance to the minimum. b) *Minimum achieved*; the algorithm finds the closest minimum to the initial point, which may be a local minimum. c) Neuron saturation; large values of the weights can involve a neuron saturation.
- *Plateaus.* Weight update is proportional to the derivative of the error function. This derivative equals zero or a very low value on the flat parts of the error function. Therefore, weights are hardly updated.
- *Choice of the learning rate.* Too high values may involve instabilities while too low values may make the converge speed very slow. There are many algorithms which propose strategies to find an optimal learning rate. They tend to be based on the following claim: *"the value of the learning rate should be high far away from the minimum and should be low near the minimum".*
- *Early stopping.* There are different criteria to carry out the learning stopping, such as, to fix a number of epochs in advance, an error threshold, to find plateaus in the cost function, etcetera. Data are usually split into two sets: a training set and a generalisation set. The former is used to train the network, whereas the latter is used to check the behaviour of the network with patterns different from those used in the training process. The goal is to obtain a model with good generalisation properties. The generalisation error tends to decrease as the learning process is progressing until a certain epoch, in which the generalisation error starts to increase because the network is overfitting the patterns of the training data set. When this change in the tendency of the

generalisation error is observed, the learning must be stopped. This procedure is known as cross-validation. There are also different criteria to decide the rate of patterns that should be assigned to each data set: 66% of the patterns to the training set, a small percentage if there is a large amount of data available, and there are also dynamic processes to carry out the selection of these sets (Haykin, 1999; Bishop, 1995).

- *Architecture choice.* The number of hidden neurons and layers is a difficult choice, being determined in many cases by using trial-and-error procedures. Nonetheless, there also pruning and growing methods that are used to find optimal structures (Reed, 1993; Reed & Marks, 1999).

3.2 Variants of the BP Algorithm

Some variants have been proposed in order to overcome the different problems observed in BP algorithm. In this section, we will focus on four of the most classical variants, namely, *Momentum term, Silva-Almeida, Delta-Bar-Delta* and *Rprop.*

3.2.1 Momentum Term

This variant (backpropagation with momentum, BPM) is very similar to the classical backpropagation algorithm. The difference is an additional term, which provides information about the weight change in the previous epoch. Therefore, weight update is given by (Haykin, 1999):

$$\Delta w_{n+1} = -\alpha \cdot (\nabla J)_n + \mu \cdot \Delta w_n \quad (6)$$

This new term controls the speed of convergence, speeding up the process when far from the minimum, and slowing it down when close to the minimum. The momentum coefficient μ gives more or less importance to the momentum term. This algorithm shows instabilities near the minimum.

3.2.1.1 Silva-Almeida

This variant adapts the value of the learning rate depending on the distance to the minimum. This distance is evaluated through two consecutive signs of the gradient of the error function. These gradients should show the same sign far away from the minimum (the algorithm is approaching the minimum in the same direction) whereas the signs should be different near the minimum since the algorithm should be oscillating around the minimum.

Weight update is identical to that carried out by the classical *backpropagation* but taking into account that learning rate is determined by (Silva & Almeida, 1990):

$$\alpha(n)=\begin{cases}\alpha(n-1)\cdot u \Leftrightarrow (\nabla J)_n\cdot(\nabla J)_{n-1}>0\\ \alpha(n-1)\cdot d \Leftrightarrow (\nabla J)_n\cdot(\nabla J)_{n-1}<0\end{cases} \tag{7}$$

being $d<1$ and $u>1$.

Moreover, this algorithm incorporates a sort of "*pocket technique*" since if an increase in the network error is observed, then the previous weights are retrieved.

3.2.2 Delta-Bar-Delta

This variant is similar to Silva-Almeida, since the learning rate is also adapted. In this case, the adaptation is given by (Jacobs, 1988):

$$\begin{aligned}&\alpha(n)=\begin{cases}\alpha(n-1)+u \Leftrightarrow (\nabla J)_n\cdot(\delta)_{n-1}>0\\ \alpha(n-1)\cdot d \Leftrightarrow (\nabla J)_n\cdot(\delta)_{n-1}<0\end{cases}\\ &(\delta)_{n-1}=(1-\theta)\cdot(\nabla J)_{n-1}+\theta\cdot(\delta)_{n-2}\end{aligned} \tag{8}$$

provided that $0<\theta<1$.

Equation (8) shows two main differences with regard to *Silva-Almeida* method:

- The increase in the learning rate to speed up the convergence is not an exponential increase, but a linear one. Therefore, it is less likely the presence of instabilities in the algorithm due to an excessively high value of the learning rate.
- The increase or decrease of the learning rate does not depend only on two consecutive gradients, but it is carried out by comparing the gradient and a weighted average of previous gradients given by the parameter δ.

A variation in weight update is proposed in (Minai & Williams, 1990):

1. An exponentially decreasing function of δ_{ij}^{n} is used instead of increasing the learning rate by using a constant factor.
2. A momentum term is added. It is updated likewise the learning rate.

3. Maximum and minimum values are imposed for the coefficients.
4. The "pocket technique" is used in order to ensure that the minimum error is obtained.

3.2.2.1 Rprop (Resilient Backpropagation)

This algorithm also proposes an adaptation of the learning rate. Moreover, weight update is somewhat different to the other variants, and it is given by (Riedmiller & Braun, 1993):

$$\Delta w_{n+1} = -\alpha \cdot sign(\nabla J)_n \tag{9}$$

Using the sign of the gradient involves less computational burden. The learning rate is adapted as follows:

$$\alpha(n) = \begin{cases} \min[\alpha(n)\cdot u, \alpha_{\max}] \Leftrightarrow (\nabla J)_n \cdot (\nabla J)_{n-1} > 0 \\ \max[\alpha(n) + d, \alpha_{\min}] \Leftrightarrow (\nabla J)_n \cdot (\nabla J)_{n-1} < 0 \end{cases} \tag{10}$$

being $u>1$ and $d<1$.

Therefore, the learning rate can show two different values, depending on the sign of the two last gradients. The aim is to have a low value of the learning rate near the minimum and a higher value far away from the minimum, thus controlling the convergence speed. The possible values of the learning rate are limited in order to avoid either an excessively high value which can lead to instabilities or a too low value which can lead to a very slow convergence speed.

Another first order algorithm is that proposed in (Chan & Fallside, 1987). This algorithm analyses the relationship between the directions defined by the gradient of the error function and the previous weight increase. In particular, it is used the following parameter:

$$\cos(\theta(n)) = -\frac{(\nabla J)_n \cdot \Delta w_{n-1}}{\|(\nabla J)_n\| \cdot \|\Delta w_{n-1}\|} \tag{11}$$

where $\|\cdot\|$ stands for the norm. This algorithm helps in overcoming the irregular parts of the error surface. This is because if there is not an appreciable change in the direction between the previous weight increase and the search direction (the gradient of the error function with opposite sign), then it means that the search process stays in a stable situation.

This Section has shown a small illustration of the most important variants of the BP algorithm. There are many other variants. A deeper review and comparison among these algorithms can be found in (Schiffmann et al., 1994), (Moreira & Fiesler, 1995).

3.3 Other Algorithms

In spite of BP algorithm and its variants are usually chosen to train MLPs, more complex algorithms have also been proposed. These algorithms tend to show a higher computational burden and faster convergence speed than BP.

3.3.1 Conjugate Gradient Algorithms

These algorithms show a convergence speed faster than that obtained by BP, in exchange for a small increase in the computational burden (Luenberger, 1984). Weight update is given by the following expression (Bishop, 1995):

$$w_{n+1} = w_n + \alpha(n) \cdot d(n) \tag{12}$$

where $\alpha(n)$ is the learning rate and $d(n)$ stands for the search direction of the minimum. This search direction is a linear combination of the gradient of the function (with opposite sign) and the previous search direction. Therefore, $d(n+1)$ is given by:

$$\begin{aligned} &d(n+1) = -g(n+1) + \beta(n) \cdot d(n) \\ &\textit{where } \mathrm{d}(0) = -g(0) \end{aligned} \tag{13}$$

$g(k)$ is the gradient of the cost function at instant k. These gradients are computed using the procedure followed in the BP algorithm. It should be pointed out that parameter $\beta(n)$ is used to define the relationship between the two directions in weight update. There are several variants of this algorithm depending on the value of this parameter (Bishop, 1995; Luenberger, 1984). The most used variants are the following (superscript t means transposition):

Fletcher-Reeves (FR). $$\beta(n) = \frac{g^t(n) \cdot g(n)}{g^t(n-1) \cdot g(n-1)} \tag{14}$$

Polak-Ribiere (PR). $\beta(n)=\dfrac{g^{t}(n)\cdot[g(n)-g(n-1)]}{g^{t}(n-1)\cdot g(n-1)}$ (15)

All these algorithms obtain the learning rate $\alpha(n)$ dynamically, using a line search procedure (Luenberger, 1984).

3.3.2 BFGS Algorithm

The goal of the learning of a neural network can be stated as a function minimisation problem. One of the most known methods of function minimisation is the Newton's method, which is faster than the other methods previously depicted. Weight update is given by the following expression:

$$w_{n+1}=w_n-[H(n)]^{-1}\cdot g(n) \quad (16)$$

The main problem of this algorithm lies in the requirement of knowing the *Hessian matrix H(n)*. This matrix contains the second derivatives of the cost function with respect to the parameters of the problem, the synaptic weights in this case. An approximation can be calculated, with an additional problem, which is the requirement of the approximation to be positive definite in order to guarantee algorithm convergence (Luenberger, 1984; Press et al., 1992). Usually, far away from the minimum and if the cost function is not quadratic, the matrix is not positive definite.

There is a kind of algorithms based on the Newton's method, the so-called quasi-Newton methods. These methods estimate the inverse Hessian matrix, forcing this matrix to be positive definite in every step (Press et al., 1992). Weight update is stated as follows:

$$w_{n+1}=w_n-\alpha(n)\cdot M(n)\cdot g(n) \quad (17)$$

M(n) is an estimation of the inverse Hessian matrix, which is updated every iteration. The most widely used quasi-Newton method is BFGS, acronym of the names of the authors who proposed the method (Broyden-Fletcher-Goldfarb-Shanno). This algorithm carries out the following processing (Luenberger 1984):

1) Algorithm initialisation; *M(0)* is considered to be any positive definite matrix, the identity matrix, for instance.
2) Search direction is calculated as follows:

$$d(n) = -M(n) \cdot g(n) \tag{18}$$

3) $\alpha(n)$ is optimised within the cost function $J[w+\alpha(n) \cdot d(n)]$ by means of a line search procedure (Luenberger, 1984; Press et al., 1992).
4) The following expressions are calculated:

$$p(n) = \alpha(n) \cdot d(n)$$
$$q(n) = g(n+1) - g(n) \tag{19}$$
$$M(n+1) = M(n) + \frac{p(n) \cdot [p(n)]^t}{[p(n)]^t \cdot p(n)} - \frac{M(n) \cdot q(n) \cdot [q(n)]^t \cdot M(n)}{[q(n)]^t \cdot M(n) \cdot q(n)}$$

5) Go to next iteration $n=n+1$, and go to step 2.

3.3.3 Levenberg-Marquardt

The Levenberg-Marquardt (LM) algorithm was proposed to be used with the quadratic cost function (Bishop, 1995):

$$J = \frac{1}{2} \cdot \sum_k (e_k)^2 \tag{20}$$

being e_k the error for pattern k and appearing the factor ½ for the sake of simplicity (after differentiation, all the constant terms are cancelled). If the vector of components e_k is denoted by $\boldsymbol{e}$, and if small perturbations of the synaptic weights are considered (Bishop, 1995):

$$\mathbf{e}\big|_{new} = \mathbf{e}\big|_{old} + L \cdot \left[w\big|_{old} - w\big|_{new} \right] \tag{21}$$

where L is a the following matrix

$$L\big|_{st} = \frac{\partial e_s}{\partial w_t} \tag{22}$$

The cost function shown in Eq. (20) can be written as:

$$J = \frac{1}{2} \cdot \sum (e_k)^2 = \frac{1}{2} \cdot \|\mathbf{e}\|^2 = \frac{1}{2} \cdot \left\| \mathbf{e}|_{old} + L \cdot \left[w|_{old} - w|_{new} \right] \right\|^2 \quad (23)$$

The minimisation of this function with respect to the new weights leads to the following weight update:

$$w|_{new} = w|_{old} - \left(L^t \cdot L \right)^{-1} \cdot L^t \cdot \mathbf{e}|_{old} \quad (24)$$

The matrix L is easy to be obtained since it only needs the first derivatives of the cost function. This procedure depends on the requirement of small changes of the synaptic weights in Eq. (21), so that if this condition is not true, then the algorithm can become instable. LM algorithm solves this problem forcing the change of the weights to be small by means of a cost function:

$$J = \frac{1}{2} \cdot \left\| \mathbf{e}|_{old} + L \cdot \left[w|_{new} - w|_{old} \right] \right\|^2 + \lambda \cdot \left\| w|_{new} - w|_{old} \right\|^2 \quad (25)$$

The minimisation of Eq. (25) with respect to the new weights leads to:

$$w|_{new} = w|_{old} - \left(L^t \cdot L + \lambda \cdot I \right)^{-1} \cdot L^t \cdot \mathbf{e}|_{old} \quad (26)$$

being I the identity matrix. If λ takes a very high value, then the BP algorithm is obtained. There are iterative procedures to obtain this factor (Bishop, 1995).

The main drawback of this algorithm lies in the need of saving the inverse matrix of L. The size of this matrix is the square of the number of synaptic weights of the network. Therefore, this method is not a good choice when the number of weights is large (Nelles, 2001).

4. EXPERIMENTAL RESULTS

In this section, the algorithms presented in previous sections are benchmarked in two different kind of problems: a classification problem and three modelling problems. Algorithms are compared in terms of accuracy and convergence speed. The accuracy achieved by the algorithm is obviously important since it is a measure of the capability of the algorithm to solve the problem. Moreover, in many practical applications (channel equalisation, for instance), the necessary time to find the solution can be almost as important as the accuracy achieved. Experiments were carried out using the hyperbolic tangent as activation function in the hidden neurons. One hundred tests were run for every experiment, using different weight initialisations; however, the same one hundred different initialisations were used in all the networks in order to obtain unbiased results. The value for the learning rate was equal to *0.5/N*, being *N* the number of hidden neurons, and the value of the momentum term was chosen as equal to *0.8*.

4.1 Classification Problems

Channel equalisation is a typical problem of application of neural networks in Communications (Qureshi, 1985; Chen et al., 1990). A general communication system is shown in Figure 1-8.

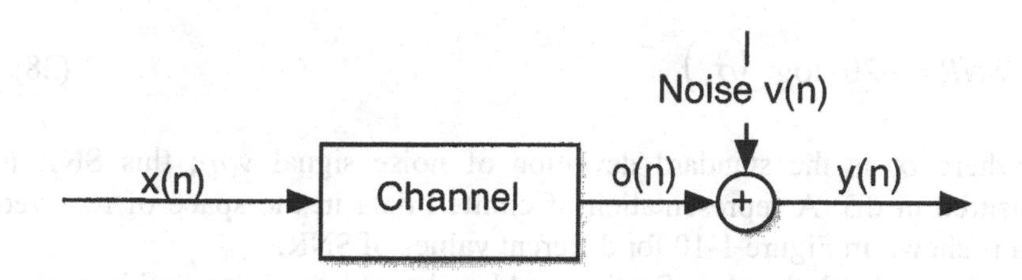

Figure 1-8. General schematic of a communication system.

Figure 1-8 shows a message *x(n)* which is sent through a communication channel. This message is modified and corrupted by the transmission channel and by the ambient noise, *v(n),* which is modelled by a Gaussian distribution with mean 0, being used its variance to characterise the noise. The goal is to decode the emitted message from the received message, as it is shown in Figure 1-9.

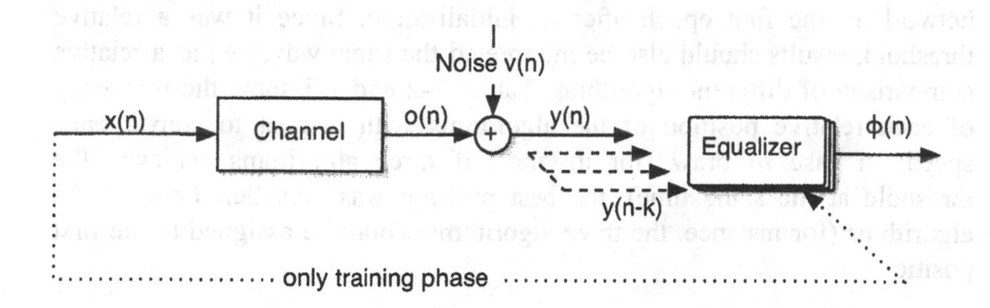

Figure 1-9. Schematic of a communication channel with equaliser.

Figure 1-9 shows how the equaliser works. During the training process, a sequence is transmitted, being this sequence known by both the emitter and the receiver (dotted line). This way, the desired signal needed by every supervised system is available. The emitted message is decoded by the equaliser, using as inputs the message received in the current instant, and also that received in previous instants, denoted as $y(n\text{-}k)$ in Figure 1-9.

We chose a widely used channel, whose difference equation is the following (Gibson et al., 1991):

$$o(n) = 0.5 \cdot x(n) + x(n-1) \tag{27}$$

The message emitted in our simulations was ±1 (this signal is known in Engineering as 2-PAM), with equal probabilities. The noise variance was varied. Signal-to-Noise Ratio (SNR) was used to characterise the transmitted signal and the environmental noise (Proakis, 2001):

$$SNR = -20 \cdot \log_{10}(\sigma_v) \tag{28}$$

where σ_v is the standard deviation of noise signal $v(n)$; this SNR is measured in dB. A representation of emitted data in the space of received data is shown in Figure 1-10 for different values of SNR.

In Figure 1-10, the classification problem is not linearly separable, and its structure is similar to other standard classification problems, as those shown in (Ripley, 1996). The main advantage of using this problem lies in the easiness to change the problem conditions, and also in their actual practical application.

Convergence speed for different algorithms is compared in Tables 1-2 and 1-3. Two different architectures were considered, setting the SNR equal to 20 dB. In order to measure convergence speed, the algorithm was supposed to converge in that iteration in which, the mean square error (MSE) of the neural network was less or equal to 15% of the error of the neural network in the first epoch after its initialisation. Since it was a relative threshold, results should also be interpreted the same way, i.e., as a relative comparison of different algorithms. Tables 1-2 and 1-3 show the frequency of each relative position of the algorithms with respect to convergence speed; in case of draw (for instance, if three algorithms achieved the threshold at the same time), the best position was considered for all the algorithms (for instance, the three algorithms would be assigned to the first position).

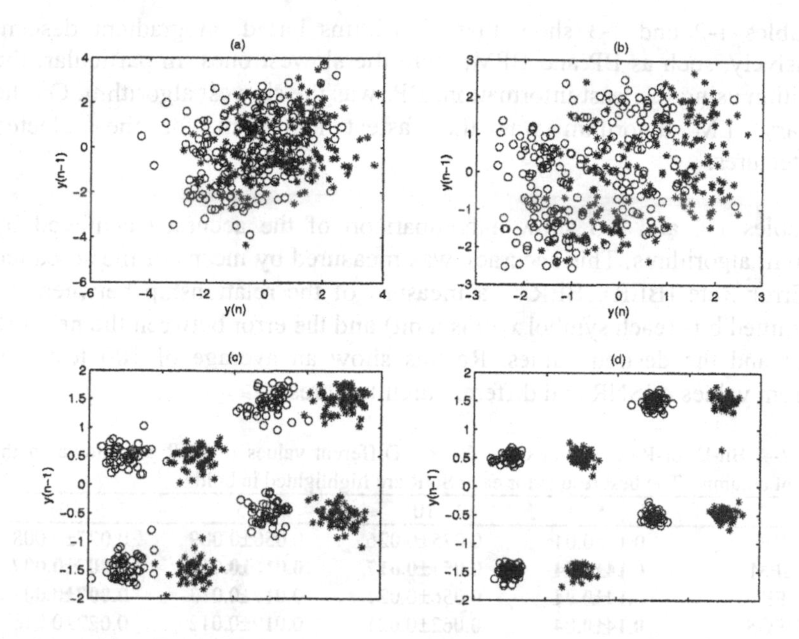

Figure 1-10. Representation of emitted data in the space of received data for different values of SNR. The transmitted symbols are '*' $x(n)=1$ and 'o' $x(n)=-1$. (a) SNR=5 dB, (b) SNR=10 dB, (c) SNR=15 dB, (d) SNR=20 dB.

Table 1-2. Architecture 2×3×2×1. Frequency of relative positions of the different algorithms in terms of convergence speed. The most frequent algorithm for each relative position is highlighted in bold

	1º	2º	3º	4º	5º
BP	0	0	1	7	**92**
BPM	0	1	18	**79**	2
FR	6	27	**56**	11	0
BFGS	32	**52**	11	4	1
LM	**64**	21	12	1	2

Table 1-3. Architecture 2×10×1. Frequency of relative positions of the different algorithms in terms of convergence speed. The most frequent algorithm for each relative position is highlighted in bold

	1º	2º	3º	4º	5º
BP	0	0	0	6	**94**
BPM	2	4	23	**68**	3
FR	9	28	**41**	21	1
BFGS	31	**52**	11	4	2
LM	**66**	17	16	1	0

Tables 1-2 and 1-3 show that algorithms based on gradient descent exclusively, such as BP and BPM, were the slowest ones. In particular, the algorithm using the least information, BP, was the slowest algorithm. On the contrary, LM algorithm was the fastest algorithm for the selected architectures.

Tables 1-4 and 1-5 show a comparison of the accuracy achieved by different algorithms. This accuracy was measured by means of the so-called Bit Error Rate (BER). BER is a measure of the relationship between the transmitted bits (each symbol *x(n)* is a bit) and the error between the network output and the desired values. Results show an average of 100 tests for different values of SNR and different architectures.

Table 1-4. Bit-Error-Rate. Architecture 2×4×1. Different values of SNR are shown in the different columns. The best results for each SNR are highlighted in bold

	5	10	15	20
BP	0.15±0.04	0.075±0.026	0.050±0.009	0.027±0.008
BPM	**0.14±0.04**	**0.054±0.017**	**0.016±0.010**	**0.007±0.007**
FR	**0.14±0.04**	0.056±0.021	0.017±0.013	**0.007±0.007**
BFGS	**0.14±0.04**	0.062±0.021	0.019±0.012	0.022±0.018
LM	**0.14±0.04**	0.056±0.018	0.031±0.016	0.028±0.015

Table 1-5. Bit-Error-Rate. Architecture 2×6×1. Different values of SNR are shown in the different columns. The best results for each SNR are highlighted in bold

	5	10	15	20
BP	0.14±0.04	0.065±0.023	0.020±0.013	0.013±0.011
BPM	**0.13±0.04**	**0.040±0.012**	**0.007±0.005**	**0.001±0.003**
FR	0.14±0.04	**0.040±0.011**	0.025±0.018	0.009±0.010
BFGS	0.14±0.04	0.046±0.014	0.013±0.007	0.021±0.015
LM	0.14±0.04	0.048±0.015	0.026±0.034	0.023±0.022

Although BPM was not the fastest algorithm, Tables 1-4 and 1-5 show that BPM did yield the most accurate results.

4.2 Modelling Problems

Three different function modelling problems were used to carry out a comparison of algorithms' modelling capabilities. First, two simple functions (f_1 and f_2) were used (Sexton et al., 2004); in addition, a more complex function was also used (f_3):

$$\begin{cases} f_1(x_1, x_2) = x_1 + x_2 \\ f_2(x_1, x_2) = x_1 \cdot x_2 \\ f_3(x_1, x_2) = \sin c(x_1) \cdot \sin c(x_2) \end{cases} \tag{29}$$

where $\sin c(x) = \dfrac{\sin(\pi \cdot x)}{\pi \cdot x}$.

Table 1-6. Problem x_1+x_2. Architecture 2×2×1. Frequency of relative positions of the different algorithms in terms of convergence speed. The most frequent algorithm for each relative position is highlighted in bold

	1°	2°	3°	4°	5°
BP	1	4	2	41	**52**
BPM	1	0	0	**54**	45
FR	29	28	**43**	0	0
BFGS	21	**61**	17	1	0
LM	**85**	6	7	1	1

Two hundred patterns formed by pairs (x_1,x_2) uniformly distributed in the range (-1,1) were used as training data set. Every experiment was carried out 100 times with the same initialisation for every neural network. As in the case of classification problems, convergence speed and accuracy was used to benchmark algorithms' performance. With respect to convergence speed, the convergence threshold was taken as a 15% of the initial MSE. Results are shown in Tables 1-6, 1-7 and 1-8.

Table 1-7. Problem $x_1 \cdot x_2$ Architecture 2×4×1. Frequency of relative positions of the different algorithms in terms of convergence speed. The most frequent algorithm for each relative position is highlighted in bold

	1°	2°	3°	4°	5°
BP	10	3	11	**43**	33
BPM	10	0	1	38	**51**
FR	**68**	23	9	0	0
BFGS	57	**31**	11	0	1
LM	61	7	**22**	3	7

Table 1-8. Problem $\text{sinc}(x_1)\cdot\text{sinc}(x_2)$. Architecture 2×4×3×1. Frequency of relative positions of the different algorithms in terms of convergence speed. The most frequent algorithm for each relative position is highlighted in bold

	1°	2°	3°	4°	5°
BP	33	5	10	22	**30**
BPM	33	6	12	**38**	11
FR	**49**	22	**29**	0	0
BFGS	59	5	11	8	17
LM	42	**37**	10	6	5

Note that LM is not the best choice always (Nelles, 2001); in fact, FR algorithm shows a faster convergence speed than LM in Tables 1-7 and 1-8.

In order to test the accuracy, the error of the neural network after training for a data set different from the training set, was measured. Five thousand patterns uniformly distributed within the range (-1, 1) were used for this accuracy test. Average results of 100 tests are shown in Table 1-9.

Table 1-9. Accuracy yielded (MSE) by the different algorithms in the three proposed problems. The best result for each problem is highlighted in bold

	x_1+x_2 2x2x1; 500 epochs	$x_1 \cdot x_2$ 2x4x1; 1000 epochs	$\mathrm{sinc}(x_1)\cdot\mathrm{sinc}(x_2)$ 2x4x3x1; 1500 epochs
BP	$(8.12\pm1.26)\cdot10^{-3}$	$(12.15\pm1.34)\cdot10^{-3}$	$(19.83\pm3.37)\cdot10^{-3}$
BPM	$(1.01\pm0.09)\cdot10^{-3}$	$(2.15\pm0.32)\cdot10^{-3}$	$(1.82\pm0.39)\cdot10^{-3}$
FR	$(8.55\pm1.67)\cdot10^{-5}$	$(6.37\pm1.41)\cdot10^{-5}$	$(1.45\pm0.21)\cdot10^{-4}$
BFGS	$(1.17\pm0.09)\cdot10^{-6}$	$\mathbf{(6.63\pm3.03)\cdot10^{-8}}$	$\mathbf{(2.41\pm0.84)\cdot10^{-5}}$
LM	$\mathbf{(1.30\pm0.67)\cdot10^{-8}}$	$(5.89\pm11.79)\cdot10^{-4}$	$(3.11\pm4.54)\cdot10^{-4}$

Table 1-9 shows that the fastest algorithms were also the most accurate ones. Nevertheless, results achieved by other algorithms were also quite accurate, hence they could also be a good choice in applications involving a high computational burden; for instance, applications involving a large number of inputs (high dimensionality).

5. CONCLUSIONS

A review of classical training methods has been provided in this chapter. It is mainly focused on the most widely used neural model, the so-called Multilayer Perceptron. It shows many attractive features; e.g., it is a universal function approximator and it is able to carry out non-linear classification. Classical training algorithms, based on the first and second derivatives, have been described. Several experiments applied to standard problems have been used to benchmark the capabilities of the different training algorithms.

Unfortunately, the reduced length of this chapter does not allow to summarise all the learning algorithms, new applications, theoretical developments and research directions related to neural models. Readers are encouraged to consult the excellent texts provided in the bibliography, as well as the following chapters of this book, as a nice way to get involved with the fascinating world of neural networks. More expert readers, rather concerned about last advances in this field, are encouraged to have a look at the updated issues of some excellent journals (*IEEE Transactions on Neural Networks*, *Neural Networks* and *Neurocomputing*, among others).

The practical implementation of the algorithms presented in this chapter is far less difficult than some years ago was. There are different software solutions for Statistics, e.g., SPSS® and Statistica®, which provide a neural network toolbox. Moreover, other software products used for numerical and symbolic computing, e.g. Matlab® and Mathematica®, also include a neural network toolbox. In addition, there is another software product (Neurosolutions®), which includes many different neural network implementations, thus allowing their use in a straightforward way.

REFERENCES

Arbib, M., 2003, *The Handbook of Brain Theory and Neural Networks*, MIT Press.

Bishop, C. M., 1995, *Neural Networks for Pattern Recognition*, Clarendon Press.

Chan, L. W., Fallside, F., 1987, An adaptive training algorithm for backpropagation, *Computer Speech and Language*, **2**:205-218.

Chen, S., Gibson, G. J., Cowan, C. F. N., Grant, P. M., 1990, Adaptive equalization of finite non-linear channels using multilayer perceptrons, *Signal Processing* **20**:107-119.

Cichocki, A., Amari, S., 2002, *Adaptive Blind Signal and Image Processing*, John Wiley & Sons.

Cybenko, G., 1988, Continuous valued neural network with two hidden layers are sufficient, Tech. Report, Department of Computer Science, Tufts University, Medford, USA.

Duda, R. O., Hart, P. E., Stork, D. G., 2001, *Pattern Classification*, Wiley.

Gibson, G. J., Siu, S., Cowan, C. F. N., 1991, The application of nonlinear structures to the reconstruction of binary signals, *IEEE Transactions on Signal Processing* **39**:1877-1881.

Haykin, S., 1996, *Adaptive Filter Theory*, Prentice-Hall.

Haykin, S., 1999, *Neural Networks: A Comprehensive Foundation*, Prentice Hall.

Hassoun, M. H.,1995, *Fundamentals of Artificial Neural Networks*, MIT Press.

Hecht-Nielsen, R., 1989, *Neurocomputing*, Addison-Wesley.

Jacobs, R. A., 1988, Increased rates of convergence through learning rate adaptation, *Neural Networks* **1**:295-307.

LeCun, Y., 1985, Une procedure d'apprentissage pour reseau a seuil asymmetrique (a Learning Scheme for Asymmetric Threshold Networks), *Proceedings of Cognitiva* **85**:599-604.

Luenberger, D. G., 1984, *Linear and Nonlinear Programming*, Addison-Wesley.

MacKay, D. J. C., 2003, *Information Theory, Inference and Learning Algorithms*, Cambridge University Press.

Minai, A. A., Williams, R. D., 1990, Acceleration of backpropagation through learning rate and momentum adaptation, in: *Proceedings of IJNCNN-90*, pp. 676-679.

Moreira, M., Fiesler, E., 1995, Neural networks with adaptive learning rate and momentum terms, Technical Report 95-04, IDIAP, Martigny, Switzerland.

Nelles, O., 2001, *Nonlinear System Identification From Classical Approaches to Neural Networks and Fuzzy Models*, Springer.

Orr, G. B., Müller, K. R., 1998, *Neural Networks: Tricks of the Trade*, Lecture Notes in Computer Science, Springer.

Press, W. H., Flannery, B. P., Teukolsky, S. A., Vetterling, W. T., 1992, *Numerical Recipes in C*, Cambridge University Press.

Proakis, J. G., 2001, *Digital Communications*, McGraw-Hill.

Qureshi, S. U. H., 1985, Adaptive equalization, *Procs. of the IEEE* **73**:1349 1387.

Reed, R., 1993, Pruning Algorithms: A Survey, *IEEE Transactions on Neural Networks* **4**(5):740-747.

Reed, R. D., Marks II, R. J., 1993, *Neural Smithing, Supervised Learning in Feedforward Artificial Neural Networks*, MIT Press.

Riedmiller, M., Braun, H., 1993, A direct adaptive method for faster backpropagation learning: the RPROP algorithm, in: *Proceedings of IEEE International Conference on Neural Networks*, pp. 586-591.

Ripley, B. D., 1996, *Pattern Recognition and Neural Networks*, Cambridge University Press.

Rumelhart, D. E., Hinton, G. E., Williams, R. J., 1986, Learning representations by back-propagating errors, *Nature* **323**:533-536.

Schiffmann, W., Joost, M., Werner, R., 1994, Optimization of the backpropagation algorithm for training multilayer perceptrons, Technical Report, University of Koblenz, Institute of Physics, Germany.

Sexton, R. S., Dorsey, R. E., Sikander, N. A., 2004, Simultaneous optimization of neural networks function and architecture algorithm, *Decision Support Systems* **36**:283-296.

Silva, F. M., Almeida, L. B., 1990, Acceleration techniques for the backpropagation algorithm, in: *Proceedings of the EURASIP Workshop*, Lecture Notes in Computer Science, vol. 412 of Lecture Notes on Computer Science, Springer-Verlag, pp. 110-119.

Weigend, A. S., Gershenfeld, N. A., 1993, *Time Series Prediction: Forecasting the Future and Understanding the Past*, Addison-Wesley.

Werbos, P. J., 1974, *Beyond Regression: New Tools for Prediction and Analysis in the Behavioral Sciences*, PhD thesis, Harvard University, Cambridge, MA, USA.

Chapter 2

SIMULATED ANNEALING

E. Aarts, P. van der Horn, J. Korst, W. Michiels, and H. Sontrop
Philips Research Laboratories, Prof. Holstlaan 4, 5656 AA Eindhoven, The Netherlands

Abstract: Simulated Annealing is a meta-heuristic that performs a randomized local search to reach near-optimal solutions of combinatorial as well as continuous optimization problems. In this chapter we show how it can be used to train artificial neural networks by examples. Experimental results indicate that good results can be obtained with little or no tuning.

Key words: Simulated annealing; neural networks.

1. INTRODUCTION

In this chapter we give an introduction to simulated annealing and show how it can be used to train artificial neural networks. Training neural networks can be seen as adapting the neurons' weights to realize a desired input-output behavior.

We can distinguish two types of artificial neural networks, namely feed-forward and recurrent networks. A feed-forward network is characterized by a partial ordering < on the nodes (neurons), such that the value of node i can only influence the value of node j, whenever $i < j$. Usually feed-forward networks are modeled to consist of multiple layers: one input layer, one or more hidden layers, and one output layer. Viewed from input to output, the value of a node in one layer only influences directly the value of the node in the next layer.

In a recurrent network the value of a given node can, via other nodes, again influence its own value. An interesting class of recurrent networks assumes bidirectional connections: if there is a connection between nodes i and j, then the value of i can influence the value of j and vice versa. Hopfield networks (Hopfield, 1982, 1984) and Boltzmann machines (Ackley, Hinton and Sejnowski, 1985) belong to this class of recurrent networks. In these networks there is no inherent difference between input and output nodes.

In this chapter, we restrict ourselves to feed-forward networks. The desired input-output behavior is characterized by a training set, i.e., a set of input-output samples. The goal of training neural networks is to adapt the nodes' weights such that after training, the neural network is able to find the corresponding output for each of the inputs in the training set. In addition, one tries to avoid overfitting, i.e., that also for inputs that are not in the training set the network is able to find the corresponding output.

Given the space limitations of this chapter, we do not discuss the use of simulated annealing in training recurrent networks. For details on how statistics on the equilibrium distribution associated with simulated annealing can be used to train Boltzmann machines, we refer to Aarts and Korst (1989).

The organization of this chapter is as follows. In Section 2 we present the basic simulated annealing algorithm. Section 3 explains how the behavior of simulated annealing can be modeled by Markov chains and gives details on its asymptotic convergence properties. Section 4 gives details on how simulated annealing can be used to train feed-forward neural networks. Implementation details and experimental results are given in Section 5.

2. BASIC SIMULATED ANNEALING ALGORITHM

In the early 1980s Kirkpatrick, Gelatt, and Vecchi (1983) and independently Cerny (1985) introduced the concepts of annealing in combinatorial optimization. In a combinatorial optimization problem we are given a finite or countably infinite set of solutions S and a cost function f that assigns a cost to each solution. The problem is to find a solution $i^* \in S$ for which $f(i^*)$ is either minimal or maximal, depending on whether the problem is a minimization or a maximization problem. Such a solution i^* is called a (globally) optimal solution. Without loss of generality, we restrict ourselves in this chapter to minimization problems.

Many interesting combinatorial optimization problems are NP-hard. For such problems it is generally believed that no algorithms exist that solve each instance in polynomial time. When confronted with an NP-hard combinatorial optimization problem, we have two options for tackling it. The first option is to aim for an optimal solution, despite the NP-hardness of the problem. A second option is to use a heuristic algorithm. Solutions found by such an algorithm are not necessarily optimal, but they are found within an acceptable amount of time. Hence, heuristic algorithms trade off optimality against computing time.

Heuristic algorithms can be classified into two categories: constructive algorithms and local search algorithms. A constructive algorithm generates a solution through a number of steps, where in each step the partial solution obtained so far is extended until in the last step a complete solution is obtained. Local search algorithms try to find high-quality solutions by searching through the solution space *S*. More specifically, a local search algorithm starts with an initial solution and then iteratively generates a new solution that is in some sense near to it. A neighborhood function $N:S \rightarrow 2^S$ defines for any given solution *s* the solutions $N(s)$ that are near to it. The set $N(s)$ is called the neighborhood of solution *s*. The process of searching through the solution space can be modeled as a walk through the (neighborhood) graph $G=(V,E)$, where node set *V* is given by the solution space *S* and arc set *E* contains arc (i,j) if and only if $j \in N(i)$.

The solution space of combinatorial optimization problems can typically be formulated in terms of discrete structures, such as sequences, permutations, graphs, and partitions. Local search uses these representations by defining neighborhood functions in terms of local rearrangements, such as moving, swapping, and replacing items, that can be applied to a representation to obtain a neighboring solution.

The simplest form of local search is iterative improvement. An iterative improvement algorithm continuously explores neighborhoods for a solution with lower cost. If such a solution is found, then the current solution is replaced by this better solution. The procedure is repeated until no better solutions can be found in the neighborhood of the current solution. By definition, iterative improvement terminates in a local optimum, which is a solution having a cost at least as good as all of its neighbors.

A disadvantage of using iterative improvement is that it easily gets trapped in poor local optima. To avoid this disadvantage, simulated annealing accepts in a limited way neighboring solutions with a cost that is worse than the cost of the current solution.

Originally the use of annealing in combinatorial optimization was heavily inspired by an analogy between the physical annealing process of solids and the problem of solving large combinatorial optimization problems. Since this analogy is quite appealing we use it here as a background for introducing simulated annealing.

In condensed matter physics, annealing is known as a thermal process for obtaining low energy states of a solid in a heat bath. The process consists of following two steps (Kirkpatrick, Gelatt, Vecchi, 1983).

- Increase the temperature of the heat bath to a maximum value at which the solid melts.
- Decrease carefully the temperature of the heat bath until the particles arrange themselves in the ground state of the solid.

In the liquid phase all particles arrange themselves randomly, whereas in the ground state of the solid, the particles are arranged in a highly structured lattice, for which the corresponding energy is minimal. The ground state of the solid is obtained only if the maximum value of the temperature is sufficiently high and the cooling is done sufficiently slow. Otherwise the solid will be frozen into a meta-stable state rather than into the true ground state.

As far back as 1953, Metropolis et al. introduced a simple algorithm for simulating the evolution of a solid in a heat bath to thermal equilibrium. Their algorithm is based on Monte Carlo techniques (Binder, 1978), and generates a sequence of states of the solid in the following way. Given a current state i of the solid with energy E_i, a subsequent state j is generated by applying a perturbation mechanism which transforms the current state into a next state by a small distortion, for instance, by displacement of a particle. The energy of the next state is E_j. If the energy difference, $E_j - E_i$, is less than or equal to 0, the state j is accepted as the current state. If the energy difference is greater than 0, the state j is accepted with a probability given by

$$\exp\left(\frac{E_i - E_j}{k_B T}\right),$$

where T denotes the temperature of the heat bath and k_B a physical constant known as the Boltzmann constant. The acceptance rule described above is known as the Metropolis criterion and the algorithm that goes with it is known as the Metropolis algorithm. It is known that, if the lowering of the temperature is done sufficiently slow, the solid can reach thermal equilibrium at each temperature. In the Metropolis algorithm this is achieved by generating a large number of transitions at a given temperature value. Thermal equilibrium is characterized by the Boltzmann distribution, which gives the probability of the solid to be in a state i with energy E_i at temperature T, and which is given by

$$\text{Prob}_T\{X = i\} = \frac{\exp(-E_i/(k_B T))}{\sum_j \exp(-E_j/(k_B T))} \tag{1}$$

where X is a random variable denoting the current state of the solid and the summation extends over all possible states. As we show below, the Boltzmann distribution plays an essential role in the analysis of the convergence of simulated annealing.

Returning to simulated annealing, the Metropolis algorithm can be used to generate a sequence of solutions of a combinatorial optimization problem by assuming the following equivalences between a physical many-particle system and a combinatorial optimization problem.

- Solutions in the combinatorial optimization problem are equivalent to states of the physical system.
- The cost of a solution is equivalent to the energy of a state.

Furthermore, we introduce a control parameter, which plays the role of the temperature. Simulated annealing can be viewed as a sequence of Metropolis algorithms, evaluated at decreasing values of the control parameter.

We now formulate simulated annealing in terms of a local search algorithm. For an instance (S,f) of a combinatorial optimization problem and a neighborhood function N, Figure 2-1 describes simulated annealing in pseudo-code.

```
procedure SIMULATED ANNEALING;
begin
  i := initial solution
  c := initial value
  repeat
    for l:=1 to L do
    begin
      probabilistically generate neighbor
      j of i
      if f(j) ≤ f(i) then accept j
      else accept j with probability
                exp((f(i) - f(j)) / c)
    end;
    update L
    update c
  until stopcriterion
end;
```

Figure 2-1. The simulated annealing algorithm in pseudo code.

The algorithm generates neighbors randomly. If a neighbor j has lower cost than the current solution i, then j is always accepted. If neighbor j has higher cost than i, then j is still accepted with a positive probability of

$$\exp\left(\frac{f(i)-f(j)}{c}\right).$$

The probability of accepting a deterioration in cost depends on the value of the control parameter c: the higher the value of the control parameter, the higher the probability of accepting the deterioration. The value of the control parameter is decreased during the execution of the algorithm. In Figure 1 the value L specifies the number of iterations that the control parameter is kept constant before it is decreased. The values of c and L and the stop criterion are specified by the 'cooling schedule'.

Initially, at large values of c, large deteriorations will be accepted; as c decreases, only smaller deteriorations will be accepted and finally, as the value of c approaches 0, no deteriorations will be accepted at all. Furthermore, there is no limitation on the size of deterioration with respect to its acceptance. In simulated annealing, arbitrarily large deteriorations are accepted with positive probability; for these deteriorations the acceptance probability is small, however. This feature means that simulated annealing, in contrast to iterative improvement, can escape from local minima while it still exhibits the favorable features of iterative improvement, namely simplicity and general applicability. The speed of convergence of simulated annealing is determined by the cooling schedule. In the next section we will indicate that under certain mild conditions on the choice of the cooling schedule simulated annealing converges asymptotically to global optima.

Comparing simulated annealing to iterative improvement it is evident that simulated annealing can be viewed as a generalization. Simulated annealing becomes identical to iterative improvement in the case where the value of the control parameter is taken equal to zero. With respect to a comparison between the performances of both algorithms we mention that for most problems simulated annealing performs better than iterative improvement, repeated for a number of different initial solutions (Fox 1994).

3. MATHEMATICAL MODELING

Simulated annealing can be mathematically modeled by means of Markov chains (Feller, 1950; Isaacson and Madsen, 1976; Seneta, 1981). In this model, we view simulated annealing as a process in which a sequence of Markov chains is generated, one for each value of the control parameter. Each chain consists of a sequence of trials, where the outcomes of the trials correspond to solutions of the problem instance.

Let (S,f) be a problem instance, N a neighborhood function, and $X(k)$ a random variable denoting the outcome of the kth trial. Then the transition probability at the kth trial for each pair $i,j \in S$ of outcomes is defined as

$$P_{ij}(k) = \text{Prob}\{X(k) = j \mid X(k-1) = i\}$$
$$= \begin{cases} G_{ij}(c_k)A_{ij}(c_k) & \text{if } i \neq j \\ 1 - \sum_{l \in S, l \neq i} G_{il}(c_k)A_{il}(c_k) & \text{if } i = j \end{cases} \quad (2)$$

where G_{ij} (c_k) denotes the generation probability, i.e., the probability of generating a solution j when being at solution i, and $A_{ij}(c_k)$ denotes the acceptance probability, i.e., the probability of accepting solution j once it is generated from solution i. The most frequently used choice for these probabilities is the following (Aarts and Korst, 1989):

$$G_{ij}(c_k) = \begin{cases} |N(i)|^{-1} & j \in N(i) \\ 0 & j \notin N(i) \end{cases} \quad (3)$$

and

$$A_{ij}(c_k) = \begin{cases} 1 & \text{if } f(j) \leq f(i) \\ \exp((f(i) - f(j))/c_k) & \text{if } f(j) > f(i) \end{cases} \quad (4)$$

For fixed values of c_k, the probabilities do not depend on k, in which case the resulting Markov chain is time-independent or homogeneous. Using the theory of Markov chains it is fairly straightforward to show that, under the condition that the neighborhoods are strongly connected - in which case the Markov chain is irreducible and aperiodic - there exist a unique stationary distribution of the outcomes. This distribution is the probability distribution of the solutions after an infinite number of trials and assumes the following form (Aarts and Korst, 1989).

Theorem 1. *Given an instance (S,f) of a combinatorial optimization problem and a suitable neighborhood function. Then, after a sufficiently large number of transitions at a fixed value c of the control parameter, applying the transition probabilities of (2), (3), and (4), simulated annealing will find a solution i∈S with a probability given by*

$$\text{Prob}_c\{X = i\} \stackrel{\text{def}}{=} q_i(c) = \frac{1}{N_0(c)} \exp\left(-\frac{f(i)}{c}\right), \quad (5)$$

where X is a stochastic variable denoting the current solution obtained by simulated annealing and

$$N_0(c) = \sum_{j \in S} \exp\left(-\frac{f(j)}{c}\right)$$

denotes a normalization constant.

A proof of this theorem is considered beyond the scope of this chapter. For those interested we refer to Aarts and Korst (1989). The probability distribution of (5) is called the stationary or equilibrium distribution and it is the equivalent of the Boltzmann distribution of (1). We can now formulate the following important result.

Corollary 1. *Given an instance (S,f) of a combinatorial optimization problem and a suitable neighborhood function. Furthermore, let the stationary distribution $q_i(c)$ that simulated annealing finds solution i after an infinite number of trials at value c of the control parameter be given by (5). Then*

$$\lim_{c \downarrow 0} q_i(c) \stackrel{\text{def}}{=} q_i^* = \frac{1}{|S^*|} C_{(S^*)}(i),$$

where S^* denotes the set of globally optimal solutions and where for any two sets A and $A' \subseteq A$ the characteristic function C is defined such that $C_{(A')}(a) = 1$ if $a \in A'$ and $C_{(A')}(a) = 0$ if $a \in A \setminus A'$.

The result of this corollary is quite interesting since it guarantees asymptotic convergence of the simulated annealing algorithm to the set of globally optimal solutions under the condition that the stationary distribution of (5) is attained at each value of c. More specifically it implies that asymptotically optimal solutions are obtained which can be expressed as

$$\lim_{c \downarrow 0} \lim_{k \to \infty} \text{Prob}_c\{X(k) \in S^*\} = 1.$$

We end this section with some remarks.

- It is possible to formulate a more general class of acceptance and generation probabilities than the ones we considered above, and prove asymptotic convergence to optimality in that case. The probabilities we used above are imposed by this more general class in a natural way and used in practically all applications reported in the literature.

- The simulated annealing algorithm can also be formulated as an inhomogeneous algorithm, viz. as a single inhomogeneous Markov chain, where the value of the control parameter c_k is decreased in between subsequent trials. In this case, asymptotic convergence again can be proved. However an additional condition on the sequence c_k of values of the control parameter is needed, namely

$$c_k \geq \frac{\Gamma}{\log(k+2)}, k = 0,1,\ldots,$$

  for some constant Γ that can be related to the neighborhood function that is applied.
- Asymptoticity estimates of the rate of convergence show that the stationary distribution of simulated annealing can only be approximated arbitrarily closely if the number of transitions is proportional to $|S|^2$. For hard problems $|S|$ is necessarily exponential in the size of the problem instance, thus, implying that approximating the asymptotic behavior arbitrarily close results in an exponential-time execution of simulated annealing. Similar results have been derived for the asymptotic convergence of the inhomogeneous algorithm.

Summarizing, simulated annealing can find optimal solutions with probability 1 if it is allowed an infinite number of transitions and it can get arbitrarily close to an optimal solution if at least an exponential amount of transitions is allowed. These results are merely of theoretical interest. The real strength of simulated annealing lies in the good practical results that can be obtained by applying more efficient finite-time implementations.

4. TRAINING NEURAL NETWORKS

In the remainder of this chapter we focus on applying simulated annealing to the problem of training a neural network. We consider a multilayer feed-forward neural network with a set V_{in} of n input nodes, a set V_{h} of m hidden nodes, and a single output node y. The neural network computes a function $h: \Re^n \rightarrow \Re$.

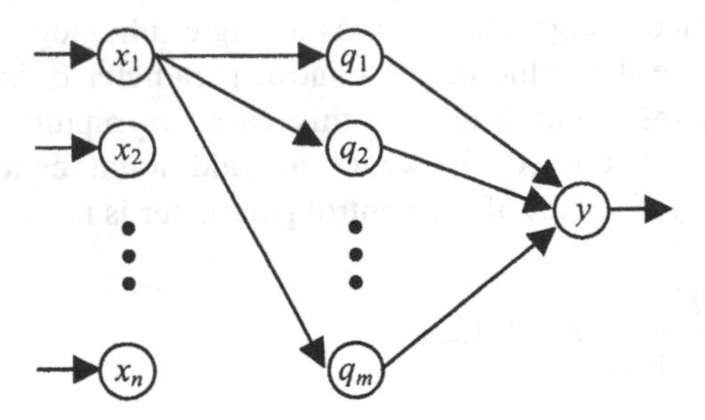

Figure 2-2. Neural network considered in this chapter. To simplify the picture, we omit the arcs from input nodes x_2 and x_n to the hidden nodes.

The neural network we consider has an arc from each input node to each hidden node and from each hidden node to the output node; see Figure 2-2. The weight of an arc from node i to node j is given by w_{ij}. Except for the input nodes, each node i has a bias value w_i. The output out_j of hidden node q_j is computed by

$$out_j = \text{sigmoid}\left(w_j + \sum_{x_i \in V_{\text{in}}} v(x_i) w_{ij} \right),$$

where sigmoid(x) is the non-linear sigmoid function $1/(1+e^{-x})$ and $v(x_i)$ is the value of input node x_i. Unlike the hidden nodes, output node y has a linear activation function. More specifically, this means that the output of node y, i.e., the output of the neural network, is computed from the outputs of the hidden layer via

$$out_y = w_y + \sum_{q_j \in V_{\text{h}}} out_j w_{jy} .$$

Suppose that we want the neural network to compute some function $g : \Re^n \to \Re$. The problem that we now face is to choose the weights in the neural network, i.e., the weights of the arcs and the biases, such that the function h computed by the neural network approximates g as good as possible. We consider the following version of this training problem.

Let $\mathcal{T}=\{T_1, T_2,..., T_r\}$ be a training set consisting of r possible inputs for the function g. Furthermore, let h_w be the function computed by the neural network of Figure 2 for weight vector w, where a weight vector specifies all the weights in the neural network. The problem is now to find a weight vector w, such that the Root Mean Square Error (RMSE) defined by

$$\sqrt{\frac{\sum_{T_i \in \mathcal{T}} error(T_i, w)}{r}}$$

is minimal, where

$$error(T_i, w) = \left(g(T_i) - h_w(T_i)\right)^2.$$

The problem of choosing the weight vector based on a training set is that the corresponding neural network may only show a good performance on the inputs in this set, but not on the entire domain of the function g that has to be learned. Therefore, after training a neural network, we do not evaluate the network by its performance on the inputs in the training set, but by its performance on the inputs in some other set, called the validation set.

5. SIMULATED ANNEALING IMPLEMENTATIONS

A weight vector for the neural network depicted in Figure 2 contains $(n+2)m+1$ weights in total: nm weights for the arcs between the input nodes and the hidden nodes, m weights for the arcs between the hidden nodes and the output node, and $m+1$ weights for the bias values. Instead of using simulated annealing to find all these weights, we only use it to find the nm weights of the arcs from the input nodes to the hidden nodes and the m weights representing the biases of the hidden nodes. For a given value of these weights, an optimal value of the other $m+1$ weights, i.e., the weights of the arcs from the hidden nodes to the output node and the bias of the output node, can be derived by means of linear regression. Given vectors X and Y, this method looks for a vector b that minimizes the squared error $e^T e$, where $e=Y-Xb$ is the error made when approximating a vector Y by the linear relation Xb. In our case, $Y=(g(T_1), g(T_2), ..., g(T_r))^T$ is the vector containing the correct function value for each input T_i from the training set $\mathcal{T}$ and X is given by

$$X = \begin{pmatrix} 1 & out_1(1) & \dots & out_m(1) \\ \vdots & \vdots & \ddots & \vdots \\ 1 & out_1(r) & \dots & out_m(r) \end{pmatrix},$$

where $out_i(j)$ defines the output of the ith hidden node for input T_j. Linear regression finds that the vector $b=(w_t,w_{1t},w_{2t},\ldots,w_{mt})$ is given by

$$b=(X^TX)^{-1}X^TY.$$

In Section 5.1 we present a basic simulated annealing implementation that is based on a neighborhood function presented by Sexton et al. (1998). This basic implementation is extended in Section 5.2.

5.1 Basic Implementation

As mentioned above we do not need to apply simulated annealing on the solution space containing all possible weight vectors. The solution space S can be restricted to weight vectors that only contain the weights of the arcs between the input nodes and the hidden nodes and the bias values of the hidden nodes. We will call such a weight vector a restricted weight vector.

Given the solution space, we can now define a neighborhood function. Let $w,v\in S$ be two restricted weight vectors. Then v is a neighbor of w with respect to some given parameter α if each weight v_i in v can be obtained from the corresponding weight w_i in w by adding or subtracting αw_i. Hence, the neighborhood N_α of w is given by

$$N_\alpha(w)=\{((1+z_1)w_1,\ldots,(1+z_{(n+1)m})w_{(n+1)m})\mid z_1,\ldots,z_{(n+1)m}\in\{-\alpha,\alpha\}\}.$$

5.2 Extended Implementation

We extend our basic implementation by introducing an additional level of local search. We do this by an approach called iterated local search (Lourenço et al., 2002). The approach works as follows. Let N_1 and N_2 be two neighborhood functions. Neighborhood function N_1 is based on the same solution space used by our basic implementation, i.e., on the set of restricted weight vectors, and neighborhood function N_2 is based on the set of solutions that are locally optimal with respect to N_1. We now search for a weight vector that minimizes RMSE by performing simulated annealing on neighborhood function N_2, where we use iterative improvement based on N_1 to derive an N_2-neighbor.

To be more specific, we define the neighborhood function N_1 as N_α for some parameter α and N_2 as $N_{\alpha\delta}$ for some additional parameter δ, where $N_{\alpha\delta}$ is defined as follows. Solution s' is an $N_{\alpha\delta}$-neighbor of s if and only if s' can be obtained from s by performing an N_δ-move followed by an execution of iterative improvement based on N_α. The parameter δ is chosen larger than

α. The idea of the resulting iterated local search algorithm is that it tries to find high-quality N_α-local optima by initializing iterative improvement with a (hopefully) high-quality starting solution that is obtained by 'kicking' a N_α-local optimum out of local optimality.

The problem of the proposed algorithm is the running time of the iterative improvement algorithm. Because $|N_\alpha|=2^{(n+1)m}$, it takes exponential time to evaluate all neighbors of a solution. Therefore, we replace the iterative improvement algorithm by a heuristic variant of the algorithm. The heuristic algorithm randomly generates neighbors of the current solution. If a solution is found with better cost, then it replaces the current solution. If some predefined number M of neighbors has been generated without finding a better solution, then the algorithm assumes that the solution is locally optimal (although this need not be the case). The resulting algorithm is depicted in Figure 2-3.

```
procedure ITERATED LOCAL SEARCH;
begin
  i := initial solution
  c := initial value
  repeat
    for l:=1 to L do
    begin
     j := 'kick' i
     k := heuristic iterative improvement on j
     if f(k) ≤ f(i) then accept k
     else accept k with probability
              exp((f(i) - f(k)) / c)
    end;
    update L
    update c
  until stopcriterion
end;
```

Figure 2-3. Iterated local search algorithm.

6. EXPERIMENTAL RESULTS

We have carried out some experiments to evaluate the two simulated annealing implementations discussed in Section 5. Our experiments are similar to the ones performed by Glover and Martí (2006) for tabu search (Chapter 4 in this book). This means that we trained a neural network with 9 hidden nodes, two input nodes x_1 and x_2, and one output node y. The domain of x_1 is given by [-100,100] and the domain of x_2 is given by [-10,10]. The training set consists of 200 inputs. These inputs are generated uniformly at random. The validation set consists of another 100 randomly generated inputs. For a precise definition of the 15 test functions used, we refer to Glover and Martí (2006).

In our basic simulated annealing implementation we use α=0.01 and L_k=1 for all iterations k. To initialize the starting value and the end value of c_k we perform a random walk through the neighborhood graph. We define the starting value of the control parameter such that if this value would be used, then the acceptance criterion of simulated annealing would accept 90% of the moves performed in the random walk. Analogously, we define the end value of the control parameter such that if this value would be used, then the acceptance criterion would accept only 1% of the moves from the random walk that lead to a cost deterioration. After each iteration of simulated annealing, the control parameter is updated by $c_{k+1} := q \cdot c_k$, where parameter q is chosen such that the simulated annealing algorithm terminates within the maximum available amount of time τ_{max}, where the algorithm terminates if the control parameter reaches its end value. Hence, q is chosen, such that if the average time per iteration is τ_{avg}, then the end value of c_k is reached after at most τ_{max}/τ_{avg} iterations. In our simulations, we let τ_{max} be 1 minute on a Pentium 4, 2.8 GHz, 1GB RAM.

In our extended simulated annealing implementation depicted in Figure 3 we use M=200. Furthermore, we let α and δ be related by $\delta=3\alpha$. While in our basic implementation α was kept constant at 0.01, we cool (decrease) α, and consequently also δ, in a similar way as we cool the control parameter. This resulted in better results than when α and δ were kept constant. The starting value of α is 0.1, and we derive the starting value of c_k and the end value of c_k in the same way as for the basic implementation. The time in which the algorithm has to terminate is again set at 1 minute.

The results of our simulations are depicted in Tables 2-1 and 2-2. We omit the functions Goldstein, Beal, and SixHumpCamelB. Similarly as the tabu search implementations of Glover and Martí (2006), our implementations were not able to train the neural network for these functions. As Glover and Martí (2006), we run both algorithms 20 times on each instance of the training problem. We report both the standard deviation and the average deviation.

Due to overfitting, the trained network sometimes shows a good performance in training, but a very poor performance on the validation set. To prevent that these results pollute the overall numbers, we only take runs of the algorithms into account in which the performance of the neural network on the training set does not differ too much from the performance on the validation set. Glover and Martí (2006) use the same strategy.

Table 2-1. Basic simulated annealing implementation

Function	Testing Average	Standard	Validation Average	Standard
Sexton 1	0	0	0	0
Sexton 2	0.35	0.28	0.43	0.38
Sexton 3	0.39	0.19	0.65	0.2
Sexton 4	0.52	0.35	0.58	0.38
Sexton 5	200.16	211.5	218.13	231.46
Branin	1917.46	1785.74	2372.63	2201.79
B2	0.69	0.4	0.76	0.44
Easom	0	0	0.01	0.02
Shubert	23.32	3.77	29.01	6.56
Booth	1.42	0.78	1.63	0.91
Matyas	0.24	0.43	0.26	0.44
Schwefel	0.81	0.49	1.01	0.6

Table 2-2. Extended simulated annealing implementation

Function	Testing Average	Standard	Validation Average	Standard
Sexton 1	0	0	0	0
Sexton 2	0.02	0.02	0.03	0.02
Sexton 3	1.16	0.64	2.4	0.76
Sexton 4	0.24	0.17	0.33	0.24
Sexton 5	7.41	7.93	15.56	20.04
Branin	89.45	34.11	124.02	41.38
B2	0.33	0.02	0.38	0.02
Easom	0	0	0	0
Shubert	17.46	2.01	37.68	9.38
Booth	0.3	0.1	0.38	0.16
Matyas	0.02	0.01	0.02	0.01
Schwefel	0.25	0.18	0.33	0.29

Glover and Martí (2006) ran their tabu search implementations for 10 minutes on a computer with the same characteristics as the computer we used. Although we run our simulated annealing implementations for only one minute, both algorithms outperform their Back Propagation and Extended Tabu Search implementations. The basic simulated annealing implementation shows a better performance for 8 out of the 12 test functions and the extended simulated annealing implementation shows a better performance for 10 out of the 12 test functions. The Extended Tabu Search Path Relinking algorithm discussed by Glover and Martí (2006) outperforms our algorithms for all functions, although its implementation is more involved.

7. CONCLUSIONS

Simulated annealing is a randomized local search algorithm that is generally applicable and easy to implement. In this chapter we introduced the metaheuristic and used it to train artificial neural networks. We proposed two simulated annealing implementations and evaluated them by carrying out some experiments. The experiments indicate that simulated annealing yields high-quality solutions with little or no tuning.

REFERENCES

Aarts, E. H. L., and Korst, J. H. M., 1989, *Simulated Annealing and Boltzmann Machines*, Wiley.

Ackley, D. H., Hinton, G. E., and Sejnowski, T. J., 1985, A learning algorithm for Boltzmann machines, *Cognitive Science* **9**: 147-169.

Binder, K., 1978, *Monte Carlo Methods in Statistical Physics*, Springer-Verlag.

Cerny, V., 1985, Thermodynamical approach to the traveling salesman problem: An efficient simulation algorithm, *Journal of Optimization Theory and Applications* **45**: 41-51.

Glover, F., and Martí, R., 2006, Tabu search, Chapter 4 in this book.

Hopfield, J. J., 1982, Neural networks and physical systems with emergent collective computational abilities, *Procs. of the Natl. Academy of Sciences of the USA* **79**: 2554-2558.

Hopfield, J. J., 1984, Neurons with graded response have collective computational properties like those of two-state neurons, *Proceedings of the National Academy of Sciences of the USA* **81**: 3088-3092.

Feller, W., 1950, *An Introduction to Probability Theory and Its Applications* **1**, Wiley.

Fox, B. L., 1994, Random restart versus simulated annealing, *Computers & Mathematics with Applications*, **27**:33-35.

Isaacson, D., and Madsen, R., 1976, *Markov Chains*, Wiley.

Kirkpatrick, S., Gelatt Jr., C. D., and Vecchi, M. P., 1983, Optimization by simulated annealing, *Science* **220**: 671-680.

Lourenço, H. R., Martin, O., and Stützle, T., 2002, Iterated local search, in: *Handbook of Metaheuristics*, F. Glover and G. Kochenberger, eds., Kluwer A. P., 321-353.

Seneta, E., 1981, *Non-negative matrices and Markov chains*, Springer-Verlag.

Sexton, R.S., Alidaee, B., Dorsey, E., and Johnson, J. D., 1998, Global optimization for artificial neural networks: a tabu search application, *European Journal of Operational Research* **106**: 570-584.

Chapter 3

TABU SEARCH

Fred Glover[1] and Rafael Martí[2]

[1] *Leeds School of Business, University of Colorado, Campus Box 419, Boulder, CO 80309;*
[2] *Dpto. de Estadística e Investigación Operativa, Universidad de Valencia, Dr. Moliner 50, 46100 Burjassot (Valencia) Spain*

Abstract: Tabu Search is a metaheuristic that guides a local heuristic search procedure to explore the solution space beyond local optimality. One of the main components of tabu search is its use of adaptive memory, which creates a more flexible search behavior. Memory based strategies are therefore the hallmark of tabu search approaches, founded on a quest for "integrating principles," by which alternative forms of memory are appropriately combined with effective strategies for exploiting them. In this chapter we address the problem of training multilayer feed-forward neural networks. These networks have been widely used for both prediction and classification in many different areas. Although the most popular method for training these networks is backpropagation, other optimization methods such as tabu search have been applied to solve this problem. This chapter describes two training algorithms based on the tabu search. The experimentation shows that the procedures provide high quality solutions to the training problem, and in addition consume a reasonable computational effort.

Key words: Intelligent problem solving; memory structures; adaptive memory programming.

1. INTRODUCTION

The basic form of Tabu Search (TS) is founded on ideas proposed by Fred Glover (1986). The method is based on procedures designed to cross boundaries of feasibility or local optimality, instead of treating them as barriers. Early examples of these procedures, derived from surrogate constraint methods and cutting plane approaches, systematically impose and released constraints to permit exploration of otherwise forbidden regions.

TS is based on the premise that problem solving, in order to qualify as intelligent, must incorporate adaptive memory and responsive exploration. The adaptive memory feature of TS allows the implementation of procedures that are capable of searching the solution space economically and effectively. Since local choices are guided by information collected during the search, TS contrasts with memory-less designs that heavily rely on semi-random processes that implement a form of sampling. The emphasis on responsive exploration (and hence purpose) in tabu search, whether in a deterministic or probabilistic implementation, derives from the supposition that a bad strategic choice can often yield more information than a good random choice, and therefore provides a basis for progressively improved strategies that take advantage of search history. TS can be directly applied to virtually any kind of optimization problem. We focus here on the non-linear problem with continuous variables that arises in the context of neural network training.

In the remainder of the chapter, we first introduce in Section 2 the basic tabu search methodology, and then in Section 3 describe the neural network training application we are interested in examining. Section 4 describes two TS methods that have been developed for solving this problem, one by Sexton et al. (1998) and the other by El_Fallahi et al. (2005). Section 5 is devoted to computational experiments that compare outcomes from these two methods, and shows the advantages these methods yield over a state-of-the-art variant of the classical backpropagation approach. The chapter finishes with relevant conclusions.

2. TABU SEARCH METHODOLOGY

In its best known form, tabu search can be viewed as beginning in the same way as ordinary local or neighborhood search, proceeding iteratively from one point (solution) to another until a chosen termination criterion is satisfied. Each solution x has an associated neighborhood $N(x) \subset X$, and each solution $x' \in N(x)$ is reached from x by an operation called a *move*.

We may contrast TS with a simple descent method where the goal is to minimize $f(x)$. Such a method only permits moves to neighbor solutions that improve the current objective function value and ends when no improving solutions can be found. The final x obtained by a descent method is called a local optimum, since it is at least as good as or better than all solutions in its neighborhood. The evident shortcoming of a descent method is that such a local optimum in most cases will not be a global optimum, i.e., it usually will not minimize $f(x)$ over all $x \in \mathbf{X}$.

Tabu search permits moves that deteriorate the current objective function value and selects the moves from a modified neighborhood $N^*(x)$. Short and long term memory structures are responsible for the specific composition of $N^*(x)$. In other words, the modified neighborhood is the result of maintaining a selective history of the states encountered during the search. In TS strategies based on short term considerations, $N^*(x)$ characteristically is a subset of $N(x)$, and the tabu classification serves to identify elements of $N(x)$ excluded from $N^*(x)$. In TS strategies that include longer term considerations, $N^*(x)$ may also be expanded to include solutions not ordinarily found in $N(x)$, such as solutions found and evaluated in past search, or identified as high quality neighbors of these past solutions. Characterized in this way, TS may be viewed as a dynamic neighborhood method. This means that the neighborhood of x is not a static set, but rather a set that can change according to the history of the search.

The structure of a neighborhood in tabu search differs from that used in local search in an additional manner, by embracing the types of moves used in constructive and destructive processes (where the foundations for such moves are accordingly called *constructive neighborhoods* and *destructive neighborhoods*). Such expanded uses of the neighborhood concept reinforce a fundamental perspective of TS, which is to define neighborhoods in dynamic ways that can include serial or simultaneous consideration of multiple types of moves.

TS uses attributive memory for guiding purposes (i.e., to compute $N^*(x)$). Instead of recording full solutions, attributive memory structures are based on recording attributes. This type of memory records information about solution properties (attributes) that change in moving from one solution to another. The most common attributive memory approaches are recency-based memory and frequency-based memory.

Recency-based memory is the most common memory structure used in TS implementations. As its name suggests, this memory structure keeps track of solutions attributes that have changed during the recent past. To exploit this memory, selected attributes that occur in solutions recently visited are labeled *tabu-active*, and solutions that contain tabu-active elements, or particular combinations of these attributes, are those that become tabu. This prevents certain solutions from the recent past from belonging to $N^*(x)$ and hence from being revisited. Other solutions that share such tabu-active attributes are also similarly prevented from being visited. Although the tabu classification strictly refers to solutions that are forbidden to be visited, by virtue of containing tabu-active attributes (or more generally by violating certain restriction based on these attributes), moves that lead to such solutions are also often referred to as being tabu.

Frequency-based memory provides a type of information that complements the information provided by recency-based memory, broadening the foundation for selecting preferred moves. Like recency, frequency often is weighted or decomposed into subclasses that can refer to particular subregions of the search space. Also, frequency can be integrated with recency to provide a composite structure for creating penalties and inducements that modify move evaluations.

A key element of the adaptive memory framework of tabu search is to create a balance between search *intensification* and *diversification.* Intensification strategies are based on modifying choice rules to encourage move combinations and solution features historically found good. They may also initiate a return to attractive regions to search them more thoroughly. Diversification strategies, on the other hand, seek to incorporate new attributes and attribute combinations that were not included within solutions previously generated. In one form, these strategies undertake to drive the search into regions dissimilar to those already examined. It is important to keep in mind that intensification and diversification are not mutually opposing, but are rather mutually reinforcing.

Most types of intensification strategies require a means for identifying a set of elite solutions as basis for incorporating good attributes into newly created solutions. Membership in the elite set is often determined by setting a threshold that is connected to the objective function value of the best solution found during the search. Two simple variants for elite solution selection have proved quite successful. One introduces a diversification measure to assure the solutions recorded differ from each other by a desired degree, and then erases all short term memory before resuming from the best of the recorded solutions. The other keeps a bounded length sequential list that adds a new solution at the end only if it is better than any previously seen, and the short term memory that accompanied this solution is also saved.

A degree of diversification is automatically created in TS by short term memory functions, but effective diversification is particularly supported by certain forms of longer term memory. TS diversification strategies are often based on modifying choice rules to bring attributes into the solutions that are infrequently used. Alternatively, they may introduce such attributes by periodically applying methods that assemble subsets of these attributes into candidate solutions for continuing the search, or by partially or fully restarting the solution process. Diversification strategies are particularly helpful when better solutions can be reached only by crossing barriers or "humps" in the solution space topology.

A TS process based strictly on short term strategies may allow a solution x to be visited more than once, but it is likely that the corresponding reduced neighborhood $N^*(x)$ will be different each time. The inclusion of longer term considerations effectively removes the risk of duplicating a previous neighborhood upon revisiting a solution and more generally of making choices that repeatedly visit only a limited subset of X.

An extensive description of the TS methodology can be found in Glover and Laguna (1997) and the integration of evolutionary methods with tabu search memory is treated in Laguna and Martí (2003).

2.1 Advanced Designs

There are many forms in which a simple tabu search implementation can be improved by adding long term elements. We restrict our attention to two of the most commonly used methods, namely strategic oscillation and path relinking, which constitute the core of many adaptive memory programming algorithms.

Strategic oscillation operates by orienting moves in relation to a critical level, as identified by a stage of construction or a chosen interval of functional values. Such a critical level or oscillation boundary often represents a point where the method would normally stop. Instead of stopping when this boundary is reached, however, the rules for selecting moves are modified, to permit the region defined by the critical level to be crossed. The approach then proceeds for a specified depth beyond the oscillation boundary, and turns around. The oscillation boundary again is approached and crossed, this time from the opposite direction, and the method proceeds to a new turning point (see Figure 3-1).

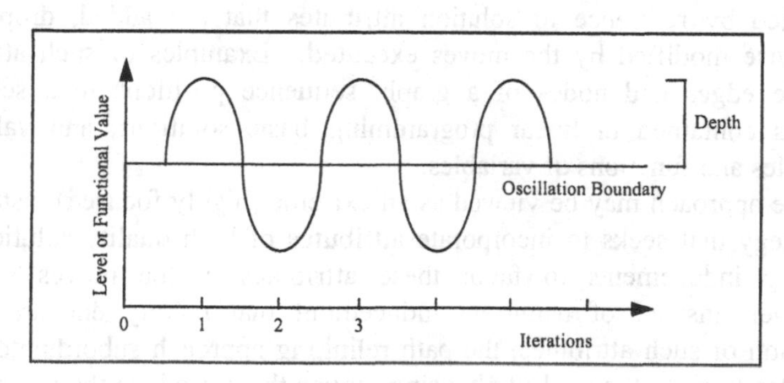

Figure 3-1. Strategic Oscillation.

The process of repeatedly approaching and crossing the critical level from different directions creates an oscillatory behavior, which gives the method its name. Control over this behavior is established by generating modified evaluations and rules of movement, depending on the region navigated and the direction of search. The possibility of retracing a prior trajectory is avoided by standard tabu search mechanisms, like those established by the recency-based and frequency-based memory functions.

When the level or functional values in Figure 1 refers to degrees of feasibility and infeasibility, a vector-valued function associated with a set of problem constraints can be used to control the oscillation. In this case, controlling the search by bounding this function can be viewed as manipulating a parameterization of the selected constraint set. A preferred alternative is often to make the function a Lagrangean or surrogate constraint penalty function, avoiding vector-valued functions and allowing tradeoffs between degrees of violation of different component constraints.

Path Relinking, as a strategy of creating trajectories of moves passing through high quality solutions was first proposed in connection with tabu search in Glover (1989). The approach was then elaborated in greater detail as a means of integrating TS intensification and diversification strategies, and given the name path relinking (PR), in Glover and Laguna (1993). PR generally operates by starting from an initiating solution, selected from a subset of high quality solutions, and generating a path in the neighborhood space that leads toward the other solutions in the subset, which are called guiding solutions. This is accomplished by selecting moves that introduce attributes contained in the guiding solutions.

Path relinking can be considered an extension of the Combination Method of Scatter Search (Glover and Laguna, 1993; Laguna and Martí, 2003). Instead of directly producing a new solution when combining two or more original solutions, PR generates paths between and beyond the selected solutions in the neighborhood space. The character of such paths is easily specified by reference to solution attributes that are added, dropped or otherwise modified by the moves executed. Examples of such attributes include edges and nodes of a graph, sequence positions in a schedule, vectors contained in linear programming basic solutions, and values of variables and functions of variables.

The approach may be viewed as an extreme (highly focused) instance of a strategy that seeks to incorporate attributes of high quality solutions, by creating inducements to favor these attributes in the moves selected. However, instead of using an inducement that merely encourages the inclusion of such attributes, the path relinking approach subordinates other considerations to the goal of choosing moves that introduce the attributes of the guiding solutions, in order to create a "good attribute composition" in the current solution. The composition at each step is determined by choosing the best move, using customary choice criteria, from a restricted set — the set of those moves currently available that incorporate a maximum number (or a maximum weighted value) of the attributes of the guiding solutions.

(Exceptions are provided by aspiration criteria, as subsequently noted.) The approach is called path relinking either by virtue of generating a new path between solutions previously linked by a series of moves executed during a search, or by generating a path between solutions previously linked to other solutions but not to each other.

To generate the desired paths, it is only necessary to select moves that perform the following role: upon starting from an initiating solution, the moves must progressively introduce attributes contributed by a guiding solution (or reduce the distance between attributes of the initiating and guiding solutions). The roles of the initiating and guiding solutions are interchangeable; each solution can also be induced to move simultaneously toward the other as a way of generating combinations. First consider the creation of paths that join two selected solutions x' and x'', restricting attention to the part of the path that lies 'between' the solutions, producing a solution sequence $x' = x(1), x(2), \ldots, x(r) = x''$. To reduce the number of options to be considered, the solution $x(i+1)$ may be created from $x(i)$ at each step by choosing a move that minimizes the number of moves remaining to reach x''. The relinked path may encounter solutions that may not be better than the initiating or guiding solution, but that provide fertile "points of access" for reaching other, somewhat better, solutions. For this reason it is valuable to examine neighboring solutions along a re-linked path, and keep track of those of high quality which may provide a starting point for launching additional searches.

3. THE NEURAL NETWORK MODEL EXAMINED

Artificial neural networks (ANNs) have been widely used for both classification and prediction. The classification or recognition problem consists of the identification of the class to which a given object belongs. The input of the net is a description of the object to be recognized, while the output is a discrete value identifying its class. The prediction problem consists in approximating unknown functions. In this chapter we restrict our attention to this latter problem; in particular we focus on the approximation of real mappings ($f: \Re^n \rightarrow \Re$).

We consider the most widely-used architecture for prediction and classification: a multilayer feed-forward network with a single hidden layer. In particular, we target a two layer feed-forward network, with sigmoid activation function in the hidden nodes and linear activation in the output node. Let $NN=(N, A)$ be an ANN where N is the set of nodes (neurons) and A is the set of arcs. N is partitioned into three subsets: N_I, input nodes, N_H, hidden nodes and N_O, output nodes. We assume that there are n variables in the function that we want to predict or approximate, therefore $|N_I| = n$. The neural network has m hidden neurons ($|N_H| = m$) with a bias term in each hidden neuron and a single output neuron. There is an arc, with an associated weight, from each node in N_I to each node in N_H, and from each node in N_H to the output node.

The net's input is given by the values of the function variables and the output is the estimation of the function image. We focus on the prediction-approximation problem, therefore, the output of the net for a given input, should be as close as possible to the value of a given function for this input. In mathematical terms, given a real function $f: \Re^n \rightarrow \Re$ and a neural net NN, the objective is to find appropriate values for the arc weights w of the net, such as its output $NN(x,w)$ from an input vector x, approximates the value $f(x)$. We refer the reader to the excellent book by Bishop (1995) for a comprehensive review of ANNs.

The most common error measure used to report the quality of the network performance is the Root Mean Squared Error (RMSE). Let $E=\{x^1, x^2, ..x^t\}$ be a random sample of points in the domain of f (usually called the training set), and suppose that the value of $f(x)$ is known for all x in E. Given the weights w, for each x in E the error can be computed as:

$$error(x,w) = [\, f(x) - NN(x,w)\,]^2,$$

and the RMSE across all the elements in the training set E is given by:

$$Error(E,w) = \sqrt{\frac{\sum_{i=1}^{t} error(x^i, w)}{t}}.$$

Therefore, training the neural network can be formulated as the following non-linear unconstrained optimization problem:

$$\min_{w} Error(E, w).$$

Backpropagation (BP) is the most widely used optimization method for accomplishing this training. It is an effective algorithm based on the steepest descent direction. Several modifications and improvements to the original method have been proposed, as it is the case of the "momentum" term where each new search direction is computed as a weighted sum of the current gradient and the previous search direction. Recently, metaheuristics such as simulated annealing (SA), genetic algorithms (GA) and tabu search (TS) have been proposed to solve this optimization problem. In this chapter we focus on the tabu search methodology and its applications to solve the neural network training problem.

4. TABU SEARCH TRAINING METHODS

In this section we describe the two tabu search implementations we will be examining for the purpose of minimizing error when training a neural network: the extended tabu search method by Sexton et al. (1998) and the tabu search approach with path relinking by El-Fallahi et al. (2005).

4.1 The Extended Tabu Search Method

The Extended Tabu Search method by Sexton et al. (1998) is mainly based on a random sampling around the best solution found. The authors propose two methods, the first one, called "preliminary" TS, is used as a baseline for comparison with the second one, Extended Tabu Search (ETS), which is the main contribution of the paper. A description of the extended method follows. Since the training set E is fixed, from now on we simplify the notation $Error(E,w)$ and use plainly $E(w)$.

An initial solution w_0 is randomly drawn from a uniform distribution in the range [-10,10] and the current best solution w_{best} is initialised to w_0. Solutions are randomly generated in this range for a given number of iterations. When generating a new point w_{new}, aspiration level and tabu conditions are checked. If $E(w_{new})<E(w_{best})$, the point is automatically accepted and both w_{best} and $E(w_{best})$ are updated; otherwise the tabu conditions are tested. If there is one solution w_i in the tabu list (TL) such as $E(w_{new}) \in [E(w_i)-0.01*E(w_i), E(w_i)+0.01*E(w_i)]$, then the complete test is applied to w_{new} and w_i; otherwise the point is accepted. The test checks if all the weights in w_{new} are within ± 0.01 from w_i, in this case the point is rejected, otherwise the point is accepted and w_{new} and $E(w_{new})$ are entered into TL. This process continues for 1000 iterations of accepted solutions. Then, another cycle of 1000 iterations of random sampling begins. These cycles will continuously repeat while $E(w_{best})$ improves.

When the random sampling ends, the process of intensification starts by performing a search from the best solution found w_{best}. The new points are drawn by modifying the w_{best} by a small *step* value, where:

$$step=((0.1*w_{best})-(0.2*w_{best})*random)/change. \qquad (1)$$

Each cycle of the intensification phase generates 1000 new points. This phase makes a maximum of 20 cycles as long as there is at least one reduction in the $E(w_{best})$. Once this phase finishes, the diversification process begins in order to expand the search area. The step value is now computed as:

$$step=((0.1*w_{best})-(0.2*w_{best})*random)*change \qquad (2)$$

This diversification phase generates new points by modifying w_{best} with step value (2). As in the intensification phase, cycles of 1000 iterations are performed up to a maximum of 20. Both phases, intensification and diversification, are alternated for a maximum of 5 consecutive iterations. The *random* variable is a random number drawn from a uniform distribution in the range [0, 1], the *change* variable is initialised to one, and is increased in one after each intensification phase. The whole process consists of 10 of these global iterations.

It is important to point out that in both this method and the one described in the next subsection, the search takes place only over the weights from the input to the hidden layer and the bias factor of the hidden neurons. Weights from the hidden layer to the output neuron, $w_{n+j,s}$ as well as the bias factor of node s, w_s, are obtained with linear regression to minimize the sum of squares associated with *Error(E,w)*. The advantage of this search scheme is that the number of weights that the training procedure needs to adjust is reduced by m+1. The disadvantage, on the other hand, is that the regression model needs to be solved every time any of the first $m(n+1)$ weights is changed in order to calculate the mean squared error.

4.2 The Tabu Search Method with Path Relinking

The tabu search algorithm by El-Fallahi et al. (2005) operates in a somewhat different manner. In this approach the short term memory is implemented in a beginning phase called TSProb. An iteration of TSProb begins by randomly selecting a weight from the current solution w. The probability of selecting weight w_i^t at iteration t, is proportional to the absolute value of the partial derivative of the RMSE on E with respect to w_i^t. These derivative values can be efficiently computed with the first phase of the BP method. The neighborhood consists of solutions that are reached from w^t by modifying the value of the selected weight w_i^t. Specifically, three solutions are considered with the following expression:

$$w_i^{t+1} = w_i^t + \alpha \beta w_i^t \;;\; w_j^{t+1} = w_j^t \;,\; \forall j \neq i$$

The method selects the best solution from among the three considered (given appropriate α values), and labels it as w^{t+1}. Note that the move is executed even when the error of w^{t+1} is greater than the error of w^t, thus resulting in a deterioration of the current value of the objective function. The moved weight becomes tabu-active for *TabuTenure* iterations, and therefore it cannot be selected during this time. The factor β scales the change in the selected weight according to the status of the search (reducing its value from 1 as long as the current solution is close to a local optimum). Starting from a random initial solution, the TSProb method finishes after a number of k consecutive iterations with no improvement. The search parameters have been set to the values recommended by the authors: *TabuTenure*= $n(m+1)/3$, $\alpha=(0.3, 0.5, 0.8)$, $\beta\in[0,1]$ and $k=500$.

The foregoing method is coupled with a Path Relinking phase, which is a form of TS strategy that is finding increasing use in applications. It starts with the creation of the Reference Set (*RefSet*), which contains the b elite solutions found during the application of the TSProb method. These b solutions must be different and they must be far enough apart to ensure that the BFGS improvement method (Smith and Lasdon, 1992) will converge to different final solutions. Therefore, a solution is admitted to *RefSet* if its Euclidean distance from each solution already in the set is larger than a pre-specified threshold *th_d*. The improvement method is applied to the $b/2$ best solutions in *RefSet* and the improved solutions are ordered according to quality (i.e., to their *error(E,w)* value).

At each iteration of the path relinking algorithm, the set *NewPairs* is constructed with all pairs of solutions in *RefSet* that include at least one new solution. (In the first iteration it contains $(b^2-b)/2$ pairs, but in successive iterations this number is usually significantly smaller.) For each pair (w', w'') in *NewPairs* a path is initiated from w' to w'', and the best solution found in the path is added to the set *PRSol.* Once all the pairs in NewPairs have been subjected to the path relinking method, the BFGS algorithm is applied to the best b solutions in *PRSol.* Each newly created solution is tested to determine whether it improves upon the worst solution currently in *RefSet,* in which case the new solution replaces the worst and *RefSet* is reordered. Then, if *RefSet* contains a new solution we perform another iteration of the path relinking algorithm, starting with the creation of the set NewPairs; otherwise, the algorithm terminates.

The path relinking method constructs a path to join two solutions u and v generated by the process described above. Considering the m neurons in the hidden layer in a given order, a path containing m solutions is constructed from solution u to solution v by performing moves that transform u into v. The first step creates the first solution in the path, w^1, by replacing in u the values of the weights in the arcs from the n input neurons to the first hidden neuron with their values in v. Similarly, the second step creates the solution w^2 by replacing in w^1 the values of the weights in the arcs from the n input neurons to the second hidden neuron with their values in v. The method proceeds in this way until we obtain solution w^m, which only differs from solution v in the values associated with the weights from the hidden layer to the output neuron.

0. Select the training set E and the testing set T. Normalize input and output data.

1. Build *RefSet* = { $w^{(1)}, \ldots, w^{(b)}$ } with the best solutions found with the TSProb method. Apply the improvement method (with stopping condition modified as explained below) to the best $b/2$ solutions in *RefSet*.

2. Order *RefSet* according to their objective function value such that $w^{(1)}$ is the best solution and $w^{(b)}$ the worst. Compute *E_best*=*error*($E, w^{(1)}$) and *T_best* as the minimum of *error*($T, w^{(i)}$) for $i=1,\ldots,b$. Set *T_Improve* = 0.

while (*T_Improve* < *T_Limit*) **do**

  3. Generate *NewPairs*, which consists of all pairs of solutions in *RefSet* that include at least one new solution. Make *NewSolutions* = $\varnothing$.

  for (all *NewPairs*) **do**

    4. Select the next pair ($w^{(i)}$, $w^{(j)}$) in *NewPairs*.

    5. Obtain new solutions in the path from $w^{(i)}$ to $w^{(j)}$ and add the best one to *NewSolutions*.

  end for

  6. Select the best b solutions in *NewSolutions* and apply the improvement method.

  for (each improved w) **do**

    if (w is not in *RefSet* and *error*(E, w) < *error*($E, w^{(b)}$)) **then**

      7. Make $w^{(b)} = w$ and reorder *RefSet*.

  end for

  8. Make *T_current* = the minimum of *error*($T, w^{(i)}$) for $i=1,\ldots,b$.

  if (*T_current* < *T_best*) **then**

    9. Make *T_best*=*T_current* and *T_improve*=0.

  else

    10. Make *T_improve*= *T_improve* +1.

end while

Figure 3-2. Path relinking pseudocode.

The effectiveness of adding a local search exploration from some of the generated solutions within the relinking path has been well documented (Laguna and Martí, 2003). In the context of neural network training, the application of the BFGS procedure as the improvement method is a time-consuming operation, so we have limited it to the best solution found in the path, as described above.

5. COMPUTATIONAL EXPERIMENTS

For our computational testing, we have created C implementations of: (1) an effective (state-of-the-art) variant of the classical Backpropagation method (BP), (2) the extended tabu search method, ETS, of Sexton et al. (1998) and (3) the TS procedure with Path Relinking, TSPR, of El-Fallahi et al. (2005). Figure 3-3 shows the expression of the 15 functions used to compare the performance of the 3 methods under consideration.

Backpropagation is one of the first methods for neural network training, and is the most widely used algorithm in practical applications. It is a gradient descent procedure that computes the derivatives' values in a very efficient way (from the output layer back towards the input layer), and modifies the weights according to a parameter known as 'learning rate'. The original algorithm has been modified in many ways; the most popular consists in adding a 'momentum' term (Rumelhart and McClelland, 1986) when the weights are updated. The inclusion of this term leads to significant improvements, although it introduces a second parameter in the algorithm. Jacobs (1988) suggested a different modification called the 'delta-bar-delta rule' which introduces a separate learning rate for each weight. It has been shown (Bishop, 1995) that this rule increases the convergence of the method in some cases, but does not work well in practice across different instances due to some stability problems. Several methods have been proposed to compute the learning rate. Examples are the *quickprop* method (Fahlman, 1988) and the *Rprop* method (Riedmiller and Heinrich, 1993). However, in general, these variants share the limitations associated with first derivative based methods. Our adaptation of the backpropagation algorithm (BP) has the learning rate set to 0.3 and the momentum term set to 1, and compares favorably with commercial implementations of this method, as documented by (El-Fallahi, 2002).

The training set consists of 200 observations with data randomly drawn from [-100, 100] for x_1 and [-10,10] for x_2. The validation set consists of 100 observations drawn from the same uniform distributions that were not used in the search process at all. We use one hidden layer with 9 nodes in all the experiments as it is done in previous works (El-Fallahi et al., 2005).

1.Sexton 1: $f(x) = x_1 + x_2$

2. Sexton 2: $f(x) = x_1 * x_2$

3. Sexton 3: $f(x) = \frac{x_1}{|x_2| + 1}$

4. Sexton 4: $f(x) = x_1^2 - x_2^3$

5. Sexton 5: $f(x) = x_1^3 - x_1^2$

6. Branin: $f(x) = \left(x_2 - \left(\frac{5}{4\pi^2} \right) x_1^2 + \left(\frac{5}{\pi} \right) x_1 - 6 \right)^2 + 10\left(1 - \frac{1}{8\pi}\right) \cos(x_i) + 10$

7. B2: $f(x) = x_1^2 + 2x_2^2 - 0.3\cos(3\pi x_1) - 0.4\cos(4\pi x_2) + 0.7$

8. Easom: $f(x) = -\cos(x_1)\cos(x_2)\exp\left(-\left((x_1 - \pi)^2 + (x_2 - \pi)^2\right)\right)$

9. Goldstein:

$$f(x) = \left(1 + (x_1 + x_2 + 1)^2 \left(19 - 14x_1 + 3x_1^2 - 14x_2 + 6x_1x_2 + 3x_2^2\right)\right) \left(30 + (2x_1 - 3x_2)^2 \left(18 - 32x_1 + 12x_1^2 + 48x_2 - 36x_1x_2 + 27x_2^2\right)\right)$$

10. Shubert: $f(x) = \left(\sum_{j=1}^{5} j\cos((j+1)x_1 + j) \right) \left(\sum_{j=1}^{5} j\cos((j+1)x_2 + j) \right)$

11. Beal:

$$f(x) = (1.5 - x_1 + x_1x_2)^2 + (2.25 - x_1 + x_1x_2^2)^2 + (2.625 - x_1 + x_1x_2^3)^2$$

12. Booth: $f(x) = (x_1 + 2x_2 - 7)^2 + (2x_1 + x_2 - 5)^2$

13. Matyas: $f(x) = 0.26(x_1^2 + x_2^2) - 0.48x_1x_2$

14. SixHumpCamelB: $f(x) = 4x_1^2 - 2.1x_1^4 + \frac{1}{3}x_1^6 + x_1x_2 - 4x_2^2 + 4x_2^4$

15. Schwefel: $f(x) = 418.9829n + \sum_{i=1}^{n} \left(-x_i \sin\sqrt{|x_i|} \right)$

Figure 3-3. Test Functions.

Tables 3-1 and 3-2 report, respectively, the training and validation errors obtained with the three methods in the 15 problems considered. In order to obtain statistically significant solutions, we run each method 20 times on each function and report the average and standard deviation of the 20 runs (limiting each run to ten minutes). In all the cases, we have employed the same training and validation sets. All the experiments have been performed on a Pentium IV 2.8 Ghz personal computer.

Table 3-1. Training Error across different methods

NP	BP	ETS	TSPR
1	1.60 ± 0.26	0.04 ± 0.02	0.00 ± 0.00
2	8.32 ± 4.30	1.79 ± 0.78	0.00 ± 0.00
3	1.63 ± 0.21	0.34 ±0.03	0.00 ± 0.00
4	45.52 ± 7.82	17.66 ±6	0.00 ± 0.00
5	12.62 ± 3.87	18.98 ±5.26	0.00 ± 0.00
6	13.98 ± 1.58	53.28 ±3.94	0.09 ± 0.04
7	16.09 ± 5.80	63.26 ±1.18	0.25 ± 0.00
8	0.20 ± 0.06	0.01 ±0.00	0.00 ± 0.00
9	7.35E+09±1.07E+09	3.30E+09±8.44E+07	1.37E+09±1.61E+08
10	21.40 ± 1.49	22.22± 4.12	16.14 ± 1.67
11	5.28E+06±1.34E+06	4.17E+06±1.28E+05	1.80E+06±1.36E+05
12	107.95 ± 3.01	156.12±5.57	0.01 ± 0.00
13	3.93 ± 1.97	10.13 ± 3.25	0.00 ± 0.00
14	5.58E+0 ± 6.76E+03	4.44E+04±2.48E+03	1.34E+04±8.34E+03
15	2.88 ± 0.5	527.14±3.07	0.02 ± 0.00

Table 3-2. Validation Error across different methods

NP	BP	ETS	TSPR
1	1.50 ± 0.22	0.05 ± 0.05	0.00 ± 0.00
2	7.91± 3.10	2.06 ± 0.85	0.00 ± 0.00
3	1.72 ± 0.21	0.67 ± 0.05	0.00 ± 0.00
4	48.03 ± 8.98	20.91 ± 7.15	0.00 ± 0.00
5	11.60 ± 2.68	21.43 ± 6.55	0.00 ± 0.00
6	15.09 ± 1.36	53.18 ± 4.80	0.00 ± 0.00
7	17.63 ± 5.87	61.2 ± 1.85	0.00± 0.00
8	0.20 ± 0.06	0.00 ± 0.00	0.00 ± 0.00
9	1.01E+10±1.75E+09	7.41E+09±3.59E+08	2.15E+09±4.58E+01
10	17.22 ± 2.9	25.59 ± 0.3	20.7 ± 0.64
11	3.83E+06±2.42E+05	5.89E+06±2.57E+05	3.29E+06±5.01E+05
12	112.09 ± 5.13	162.47 ± 7.34	0.00 ± 0.00
13	4.72 ± 2.74	10.52 ± 0.41	0.01 ± 0.00
14	5.20E+04±5.54E+03	4.38E+04±2.96E+03	1.44E+04±7.78E+03
15	2.78 ± 0.46	528.26±18.48	0.03 ± 0.00

Tables 3-1 and 3-2 show that the best solution quality is obtained with the TSPR method in all cases. Note that this method presents a lower average error as well as a lower deviation than the other methods. This experiment also shows that none of the methods can effectively handle problems 9, 11 and 14 within the run time considered, suggesting either that the ANN model itself needs to be modified in these cases or that the structure of these problems poses an unusual level of difficulty. Considering the average values over the 20 runs, Table 1 shows that the TSPR method is able to obtain the best solutions with respect to the training error in each of the other 12 instances. Table 2 shows similar results since TSPR obtains the best solutions in the 12 cases with reasonable error values. It should be mentioned that our BP algorithm variant obtains deviation values that are reasonably good, on average, considering its simplicity.

Figure 3-4 shows the average validation error on the 12 instances in which the methods provide relative good approximation. We have replaced the 528.26 achieved with the ETS method on instance 15 with a 180 for scaling purposes. This figure clearly shows the superiority of the TSPR method compared with the other approaches under consideration.

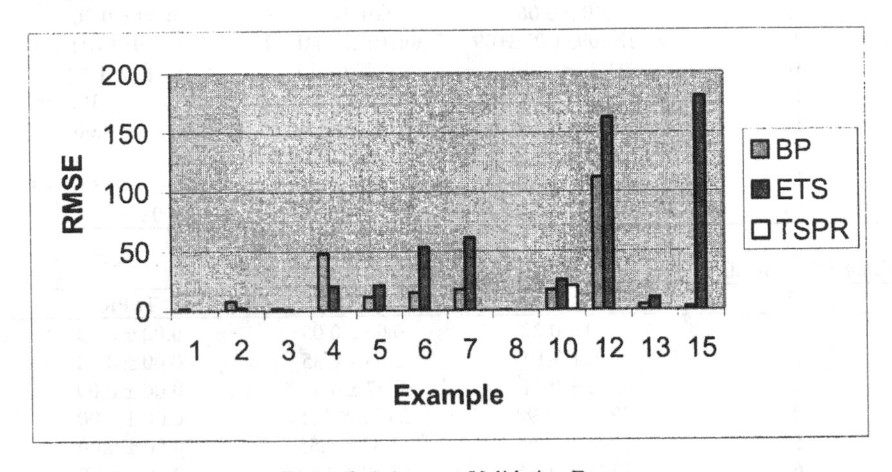

Figure 3-4. Average Validation Error.

6. CONCLUSIONS

The focus and emphasis of tabu search have a number of implications for the goal of designing improved optimization procedures. These opportunities carry with them an emphasis on producing systematic and strategically designed rules, rather than following the policy of relegating decisions to random choices as is often fashionable in evolutionary methods. The adaptive memory structures underlying tabu search and the excellent results that they provide, invites the use of TS in other metaheuristics.

We have described different implementations of tabu search for training a single-layer feed-forward neural network. Two TS methods were compared with the well known Backpropagation algorithm. The best results are obtained by the Tabu Search Path Relinking method coupled with an improvement phase based on the BFGS optimizer.

ACKNOWLEDGMENTS

Research by Rafael Martí is partially supported by the *Ministerio de Educación y Ciencia* (refs. TIN2004-20061-E and TIC2003-C05-01) and by the *Agencia Valenciana de Ciència i Tecnologia* (ref. GRUPOS03 /189).

REFERENCES

Bishop, C. M., 1995, *Neural Networks for Pattern Recognition*, Oxford University Press.

El-Fallahi A., 2002, *Entrenamiento de Redes Neuronales*, Trabajo de Investigación, Dpto Estadística e I.O. University of Valencia.

El-Fallahi, A., Martí, R., and Lasdon, L., 2005, Path relinking and GRG for artificial neural networks, *European journal of operational research* **169**:508-519.

Fahlman, S. E., 1988, An empirical study of learning speed in back-propagation networks, in: *Connectionist Models Summer School*, T. J. Sejnowski, G. E. Hinton and D. S. Touretzky, eds., San Mateo, CA, Morgan Kaufmann, pp. 38-51.

Glover, F., 1986, Future paths for integer programming and links to artificial intelligence, *Computers and Operations Research* **13**:533-549.

Glover, F., 1989, Tabu search part I, *ORSA Journal on Computing* **1**:190-206.

Glover, F., and Laguna, M., 1993, Tabu search, in: *Modern Heuristic Techniques for Combinatorial Problems,* C. Reeves, ed., Blackwell Scientific Publishing, Oxford, pp. 70-150.

Glover, F., and Laguna, M., 1997, *Tabu Search,* Kluwer Academic Publishers, Boston.

Jacobs, R. A., 1988, Increased rates of convergence through learning rate adaptation, *Neural Networks* **1**:295-307.

Laguna, M., Martí, R., 2003, *Scatter Search—Methodology and Implementations in C,* Kluwer Academic Publishers, Boston.

Riedmiller, M., and Heinrich, B., 1993, A direct adaptive method for faster back-propagation learning: the RPROP algorithm, *IEEE Intl. Conf. on Neural Networks*, pp. 586-591.

Rumelhart, D. E., and McClelland, J. L., 1986, *Parallel distributed processing: explorations in the microstructure of cognition*, Cambridge, MA: The MIT Press.

Sexton, R. S., Alidaee, B., Dorsey, R. E., and Johnson, J. D., 1998, Global optimization for artificial neural networks: a tabu search application, *European Journal of Operational Research* **106**:570-584.

Smith, S., and Lasdon, L., 1992, Solving large nonlinear programs using GRG, *ORSA Journal on Computing* **4**(1): 2-15.

Chapter 4

VARIABLE NEIGHBOURHOOD SEARCH

José Andrés Moreno Pérez[1], Nenad Mladenović[2], Belén Melián Batista[1] and Ignacio J. García del Amo[1]

[1]*Grupo de Computación Inteligente, Instituto Universitario de Desarrollo Regional, ETS Ingeniería Informática. Universidad de La Laguna, 38271 La Laguna, Spain;*

[2]*School of Mathematics, Brunel University, West London, United Kingdom*

Abstract: The basic idea of VNS is the change of neighbourhoods in the search for a better solution. VNS proceeds by a descent method to a local minimum exploring then, systematically or at random, increasingly distant neighbourhoods of this solution. Each time, one or several points within the current neighbourhood are used as initial solutions for a local descent. The method jumps from the current solution to a new one if and only if a better solution has been found. Therefore, VNS is not a trajectory following method (as Simulated Annealing or Tabu Search) and does not specify forbidden moves. In this work, we show how the variable neighbourhood search metaheuristic can be applied to train an artificial neural network. We define a set of nested neighbourhoods and follow the basic VNS scheme to carry out our experiments

Key words: Variable neighbourhood search; neural networks.

1. INTRODUCTION

Artificial neural networks constitute a tool to approximate non-linear mappings from several input variables to several output variables. In order to perform the approximation, the structure of the network has to be determined and a set of parameters, known as weights, has to be tuned. Depending on the domain of the output values, two kinds of problems can be tackled: approximation or prediction problems, for which the output values of the network are continuous variables, and classification problems, for which the output is a single categorical variable. Most of the key issues in the net functionality are common to both.

The main goal in the fitting process is to obtain a model that makes good predictions for new inputs (i.e. to provide good generalization). Once the structure of the network is given, the problem is to find the values of the weights **w** that optimize the performance of the network in the classification or prediction tasks.

In the supervised learning approach, given a training data set, the network is trained for the classification or prediction tasks by tuning the values of the weights in order to minimize the error across the training set. The training set *T* consists of a series of input patterns and their corresponding outputs. If the function *f* to be approximated or predicted has an input vector of variables $\mathbf{x} = (x_1, x_2, \ldots, x_n)$ and the output is represented by $f(\mathbf{x})$, the error of the prediction is the difference between the output $p(\mathbf{w},\mathbf{x})$ provided by the network and the real value $f(\mathbf{x})$. The usual way to measure the total error is by the root mean squared difference between the predicted output $p(\mathbf{w},\mathbf{x})$ and the actual output value $f(\mathbf{x})$ for all the elements x in T (*RMSE*; Root Mean Squared Error).

$$RMSE(T,w) = \sqrt{\frac{1}{|T|}\sum_{x \in T}(f(x) - p(w,x))^2}$$

Therefore, the task of training the net by tuning the weights is interpreted as the non-linear optimization problem of minimizing the *RMSE* on the training set through an optimal set of values w^* for the weights. This is equivalent to solve the following problem:

$$RMSE(T,w^*) = \min_{w} RMSE(T,w)$$

To solve this problem, one can apply specific or general optimization techniques. However, the main goal in the development of an artificial neural network is to obtain a design that makes good predictions for future inputs (i.e. which achieves the best possible generalization). Therefore, the design must allow the representation of the systematic aspects of the data rather than their specific details. The standard way to evaluate the generalization provided by the network consists of introducing another set of input/output pairs, *V*, in order to perform the validation. Once the training has been performed and the weights have been chosen, the performance of the design is given by the *RMSE* across the validation set *V*, i.e., the *validation error*, stated as:

$$RMSE(V, w^*) = \sqrt{\frac{1}{|V|}\sum_{y \in V}(f(y) - p(w^*, y))^2}$$

The net must exhibit a good fitting between the target values and the output (prediction) in both the training set and the testing set. If the *RMSE* in T is significantly higher than that one in V, we will say that the net has memorized the data, instead of learning them (i.e., the net has over-fitted the training data). In order to avoid over-fitting, the training process should stop before the network starts to memorize the data instead of learning the general characteristics of the instances. This can be done by introducing a third disjoint set of instances V_T, the test set, for which the *RMSE* is calculated after a certain number of training iterations. If this error increases instead of decreasing, then the training stops.

If the design of the model has too few parameters, it is difficult to fit the network to the data. On the other hand, if it has too many parameters and the structure is general enough, it would over-fit the training data, decreasing the training error and increasing both the test and validation errors. However, there is not a well-established criterion to determine the appropriate number of parameters, and it is difficult to reach a consensus on what the good architectures of the networks are.

The usual artificial neural networks used for the approximation or prediction of a function consist of a series of input nodes or neurons (one for each variable of the function) and one output neuron (or several if the function is multi-dimensional). These input-output neurons are interconnected by additional sets of hidden neurons, which number are variable and have to be determined at design time. A kind of structure that has been widely used is the multilayer architecture, where the neurons are organized in *layers* (*input*, *hidden* and *output*) with connections (weights) only between neurons of different layers. These connections are usually set in a *feed-forward* way, which means that the output of the neurons in one layer is transformed by the weights and fed to the neurons in the next layer.

The typical model consists of a network with only one hidden layer with h neurons. Therefore, the set of neurons N of the network is divided into n input neurons (n is the number of variables of the function to be approximated) in the input layer N_I, h neurons in the hidden layer N_H, and a single neuron in the output layer N_O. The connections follow the *feed-forward* model, i.e., all the connections go from the neurons in the input layer to the neurons in the hidden layer and from the neurons in the hidden layer to the neuron in the output layer.

Blum and Li (1991) proved that a neural network having two layers and sigmoid hidden units could approximate any continuous mapping arbitrarily well. As consequence, regarding to the classification problem, two layer networks with sigmoid units can approximate any decision boundary with arbitrary accuracy. However, Gallant and White (1992) showed that, from a practical point of view, the number of hidden units must grow as the size of the data set to be approximated (or classified) grows.

Within a multilayer neural network, the neurons can be enumerated consecutively through the layers from the first to the last one. So we consider a network with a hidden layer to predict a real valued function with n variables consisting of a set of input neurons $N_I = \{ 1, 2, \ldots, n \}$, a set of hidden neurons $N_H = \{ n+1, n+2, \ldots, n+h \}$ and the output neuron $n+h+1$. The links are:

$$L = \{ (i,n+j): i = 1, \ldots, n, j = 1, \ldots, h \} \cup \{ (n+j,n+h+1): j = 1, \ldots, h \}.$$

This network is shown in *Figure 4-1*.

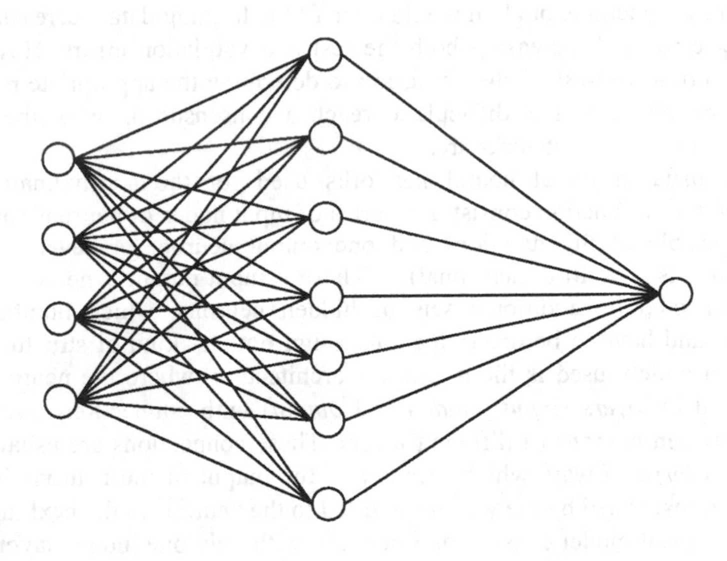

Figure 4-1. General scheme for an artificial neural network.

Given an input pattern $x = (x_1, x_2, \ldots, x_n)$ for the input neurons of the network, each hidden neuron receive an input from each input neuron to which it is connected and sends its output to the output neuron.

In the usual models, the inputs of each neuron of the hidden and output layers are linear combinations of the weights of the links and the output of the previous layer. So, the input for the j-th neuron of the hidden layer is:

$$x_j = w_j + \sum_{i=1}^{n} x_j w_{ij}; j = n+1, n+2, \ldots, n+h.$$

Here, w_j is the weight associated to the bias of the previous layer (usually, implemented as an "extra" neuron which always outputs a value of 1.0). Each neuron of the hidden layer, transform its input into an output by the expression $y_j = g(x_j)$ being g the sigmoid function $g(x) = 1 / (1 + exp(-x))$ one of the most used function. The sigmoid's output is in the range (0, 1), but other variants of this function may be used in order to adjust the output to a different range. For prediction problems, it is also usual to consider a linear activation function for the output layer.

2. THE VNS METHODOLOGY

An optimization problem consists in finding the minimum or maximum of a real valued function f defined on an arbitrary set X. If it is a minimization problem, it can be formulated as follows:

$$\min\{f(x) : x \in X\} \tag{1}$$

In this notation, X denotes the *solution space*, x represents a *feasible solution* and f the *objective function* of the problem. It is a combinatorial optimization problem if the solution space is discrete or partially discrete. An *optimal solution* x^* (or a global minimum) of the problem is a feasible solution for which the minimum of (1) is reached. Therefore, $x^* \in X$ satisfies that $f(x^*) \leq f(x)$, $\forall x \in X$. A *Neighbourhood Structure* in X is defined by a function $N: X \rightarrow 2^X$ where, $\forall x \in X$, $N(x) \subseteq X$ is the set of *neighbours* of x. Then, a *local minimum* x' of (1), with respect to (w.r.t. for short) the neighbourhood structure N, is a feasible solution $x' \in X$ that satisfies the following property: $f(x') \leq f(x)$, $\forall x \in N(x')$. Therefore any *local* or neighbourhood *search* method (i.e., method that only moves to a better neighbour of the current solution) is trapped when it reaches a local minimum.

Several metaheuristics, or frameworks for building heuristics, extend this scheme to avoid being trapped in a local optimum. The best known of them are *Genetic Search*, *Simulated Annealing* and *Tabu Search* (for discussion of these metaheuristics and others, the reader is referred to the books of surveys edited by Reeves (1993) and Glover and Kochenberger (2003)). *Variable Neighbourhood Search* (VNS) (Mladenović and Hansen (1997), Hansen and Mladenović (2001, 2003)) is a recent metaheuristic that exploits systematically the idea of neighbourhood changes with the purpose of both reaching local minima and escaping the valleys which contain them.

VNS runs a descent method from a starting point to reach a local minimum. It explores then a series of different predefined neighbourhoods of this solution. Each time one or several points of the current neighbourhood are used as starting points to run the local descent method it stops at a local minimum. The search jumps to the new local minimum if and only if it is better than the incumbent. In this sense, VNS is not a trajectory following method (that allows non-improving moves within the same neighbourhood) as Simulated Annealing or Tabu Search.

Unlike many other metaheuristics, the basic schemes of VNS and its extensions are simple and require few parameters, and sometimes none at all. Therefore, in addition to providing very good solutions, often in simpler ways than other methods, VNS gives insight into the reasons for such a performance, which in turn can lead to more efficient and sophisticated implementations. Despite its simplicity, it proves to be effective. VNS exploits systematically the following observations:

1. A local minimum with respect to one neighbourhood structure is not necessary so for another.
2. A global minimum is a local minimum with respect to all possible neighbourhood structures.
3. For many problems local minima with respect to one or several neighbourhoods are relatively close to each other.

The last observation, which is empirical, implies that a local optimum often provides some information about the global one. There may for instance be several variables with the same value in both. However, it is usually not known which ones are such. An organized study of the neighbourhoods of this local optimum is therefore performed in order, until a better one is found.

Variable Neighbourhood Descent (VND) is a deterministic version of VNS. It is based on the *observation 1* mentioned above, i.e., *a local optimum for a first type of move* $x \leftarrow x'$ (either heuristic or within the neighbourhood $N_1(x)$) *is not necessary one for another type of move* $x \leftarrow \tilde{x}$ (within neighbourhood $N_2(x)$). It may thus be advantageous to combine descent heuristics. This leads to the basic VND scheme presented in *Figure 4-2*. It is assumed that the series of neighbourhood structures N_k, $k = 1, \cdots, k_{max}$, that will be used in the algorithm are given.

VND method

1. Find an initial solution x.
2. Repeat the following sequence until no improvement is obtained:
   (i) Set $l \leftarrow 1$;
   (ii) Repeat the following steps until $l = l_{\max}$:
      (a) Find the best neighbor x' of x ($x' \in N_l(x)$);
      (b) If the solution x' thus obtained is better than x, set $x \leftarrow x'$ and $l \leftarrow 1$; otherwise, set $l \leftarrow l+1$;

Figure 4-2. Variable Neighbourhood Descent (VND).

Another simple application of the VNS principles appears in the reduced VNS. It is a pure stochastic search method: solutions from the pre-selected neighbourhoods are chosen at random. Its efficiency is mostly based on *observation 3* described above. A set of neighbourhoods $N_1(x)$, $N_2(x)$, $\cdots$, $N_{k_{max}}(x)$ around the current point x (which may be or not a local optimum) is considered. Usually, these neighbourhoods are nested, i.e., each one contains the previous. Then, a point x' is chosen at random in the first neighbourhood. If its value is better than that of the incumbent (i.e., $f(x') < f(x)$), the search is re-centred there ($x \leftarrow x'$). Otherwise, it proceeds to the next neighbourhood. After all neighbourhoods have been considered, the search begins again with the first one, until a stopping condition is satisfied (usually it is the maximum computing time since the last improvement, or the maximum number of iterations). The description of the steps of *Reduced VNS* is as shown in *Figure 4-3*. It is assumed that the neighbourhood structures N_k, $k = 1, \cdots, k_{max}$, that will be used in the shake are given.

RVNS method

1. Find an initial solution x; choose a stopping condition;
2. Repeat the following sequence until the stoping condition is met:
   (i) Set $k \leftarrow 1$;
   (ii) Repeat the following steps until $k = k_{\max}$:
      (a) *Shake.* Take at random a solution x' from $N_k(x)$;
      (b) If the solution x' is better than the incumbent, move there ($x \leftarrow x'$) and continue the search with N_1 ($k \leftarrow 1$); otherwise, set $k \leftarrow k+1$;

Figure 4-3. Reduced Variable Neighbourhood Search (RVNS).

In the two previous methods, we examined how to use variable neighbourhoods in descent to a local optimum and in finding promising regions for near-optimal solutions. Merging the tools for both tasks leads to the *General Variable Neighbourhood Search* (GVNS) scheme. We first discuss how to combine a local search with systematic changes of neighbourhoods around the local optimum found. We then obtain the *Basic VNS* scheme of *Figure 4-4*. Here also, it is assumed that that the neighbourhood structures N_k, $k = 1, \cdots, k_{max}$, that will be used in the shaking are given.

BVNS method

1. Find an initial solution x; choose a stopping condition;
2. Repeat the following sequence until the stoping condition is met:
   (i) Set $k \leftarrow 1$;
   (ii) Repeat the following steps until $k = k_{\max}$:
      (a) *Shaking.* Generate a point x' at random from the k-th neighborhood of x ($x' \in N_k(x)$);
      (b) *Local Search.* Apply some local search method with x' as initial solution; denote with x'' the so obtained local optimum;
      (c) *Move or not.* If the local optimum x'' is better than the incumbent x, move there ($x \leftarrow x''$) and continue the search with N_1 ($k \leftarrow 1$); otherwise, set $k \leftarrow k+1$;

Figure 4-4. Basic Variable Neighbourhood Search (BVNS).

The simplest Basic VNS, where the neighbourhood for shaking are fixed, is called *Fixed Neighbourhood Search* (FNS) (see Brimberg et al. (2000)) and sometimes called *Iterated Local Search*, (see Lourenco et al. (2003)). This method selects a neighbour of the current solution by a perturbation, runs a local search from it to reach a local optimum, and moves to it if there has been an improvement. Therefore, the definition of different neighbourhood structures is not necessary, as it can consider only one among them (i.e., by fixing k) and jump (or 'kick the function') in the shaking (or perturbation) step to a point from that fixed neighbourhood. For example in Johnson and Mc-Geosh (1997) a new solution is always obtained from 4-opt (double-bridge) neighbourhood in solving TSP. Thus, k is fixed to 4. The steps of the FNS are obtained considering only one neighbourhood (see *Figure 4-5*).

FNS method

1. Initialization:
   Find an initial solution x; Set $x^* \leftarrow x$;
2. Iterations:
   Repeat the following sequence until a stopping condition is met:
   (a) *Shake.*
   Take at random a neighbor x' of x ($x' \in N(x)$);
   (b) *Local Search.*
   Apply the local search method with x' as initial solution; denote x'' the so obtained local optimum;
   (c) *Move or not.*
   If x'' is better than x^*, do $x^* \leftarrow x''$

Figure 4-5. Fixed Neighbourhood Search (FNS).

If VND is used instead of simple local search, and the initial solution found by Reduced VNS is improved, the *General Variable Neighbourhood Search* scheme (GVNS) is obtained (shown in *Figure 4-6*). Here, in addition to the set of neighbourhood structures (N_k, $k = 1, \cdots, k_{max}$) to be used in the shaking, the set of neighbourhood structures (N_l, $l = 1, \cdots, l_{max}$) that will be used in the local search are given.

GVNS method

1. Initialization:
   Find an initial solution x and improve it by using RVNS.
2. Iterations:
   Repeat the following sequence until the stopping condition is met:
   (i) Set $k \leftarrow 1$;
   (ii) Repeat the following steps until $k = k_{max}$;
   (a) *Shaking.*
   Generate at random a point x' in the k-th neighborhood of x ($x' \in N_k(x)$);
   (b) *Local Search by VND.*
   Set $l \leftarrow 1$; and repeat the following steps until $l = l_{max}$;
   1. Find the best neighbor x'' of x in $N_l(x')$
   2. If $f(x'') < f(x')$ set $x' \leftarrow x''$ and $l \leftarrow 1$; otherwise set $l \leftarrow l + 1$;

   (c) *Move or not.*
   If this local optimum is better than the incumbent, move there ($x \leftarrow x''$), and continue the search with N_1 ($k \leftarrow 1$); otherwise, set $k \leftarrow k + 1$;

Figure 4-6. General Variable Neighbourhood Search (GVNS).

Also, a C code for the basic version of the Variable Neighbourhood Search is shown in *Figure 4-7.*

```
BVNS Code
 1: initialize(best_sol) ;
 2: while (t < t_max) {
 3:   k = 0 ;
 4:   while (k < k_max) {
 5:     k++ ;
 6:     cur_sol = shake(best_sol,k) ;
 7:     local_search(cur_sol,best_sol)
 8:     if improved(cur_sol,best_sol) {
 9:       best_sol = cur_sol ;
10:       k = 0 ;
11:     } /* if */
12:   } /* while k */
13: } /* while t */
```

Figure 4-7. Basic Variable Neighbourhood Search code.

This code of the VNS can be applied to any problem if the user provides the initialization procedure `initialize`, the shake procedure `shake`, the local search procedure `local_search`, and the function `improved` to test if the solution is improved or not.

3. APLICATION OF THE VNS TO THE ARTIFICIAL NEURAL NETWORKS TRAINING PROBLEM

When considering the problem of training an artificial neural network from the metaheuristics point of view, it is useful to treat it as a global optimization problem with the following considerations. The error E is a function of the adaptive parameters of the net, and we can arrange them all (i.e., weights and biases) into a single W-dimensional weight vector $\boldsymbol{w}$ with components $w_1, w_2, \cdots, w_W$. Now, the problem consists in finding the weight vector $\boldsymbol{w}^*$ with the lowest error. All over this chapter we will talk about solutions, meaning simply vectors of weights, as there are no restrictions to the values a weight can have, and thus, every point of the W-dimensional space is a feasible solution. From now on in this section, we will try to explain how a VNS variant can be applied to this problem.

The key point in the VNS is that it searches over several neighbourhood structures in the solution space. A neighbourhood of a solution s_1 is simply a group of solutions that are related to s_1 through a neighbourhood function. Usually, this function is the classical Euclidean distance, so a solution s_2 belongs to the k-th neighbourhood of a solution s_1 if the Euclidean distance between s_2 and s_1 is less than or equal to k. From this example we can also deduce that this function generates nested neighbourhoods, in the sense that if a solution s_2 belongs to the k-th neighbourhood of s_1, it also belongs to its $(k+1)$-th neighbourhood. This is a desirable property, because if we find that a certain solution s_3 is a local minimum of the k-th neighbourhood, it is also most likely a local minimum of all the previous neighbourhoods $1, \cdots, k-1$.

If the solution space has no inner structure, or this structure is unknown to us, the Euclidean distance is a neighbourhood function as good as any other can be, with the advantage that it is an intuitive function and generates nested neighbourhoods. However, if the solution space does have some kind of inner structure and some information about it is known, it can be used to generate a neighbourhood function that is better fitted to the problem. In the artificial neural network training problem, it seems obvious that the solutions do have such a structure. In fact, we know the structure, as the elements of a solution (the weights) have to be arranged spatially according to the architecture of the net. Then, the goal is to use this information to generate a good neighbourhood function.

With the purpose of determining a possible neighbourhood function that allows us to define a set of nested neighbourhoods, we use some ideas proposed by El Fallahi et al. (2005). The basic idea consists in sorting the neurons in descending order according to the partial errors of their incoming weights respect to the global network error. This can be done by making use of the classical *Backpropagation* method (Rumelhart et al., 1986) to calculate the contribution of each weight to the global error of the network. Then, each neuron is assigned a value equal to the error ratio of its incoming weights. Lastly, we define the k-th neighbourhood of a solution as all the solutions that can be reached by changing the incoming weights of the k first neurons in the ordering (see *Figure 4-8*).

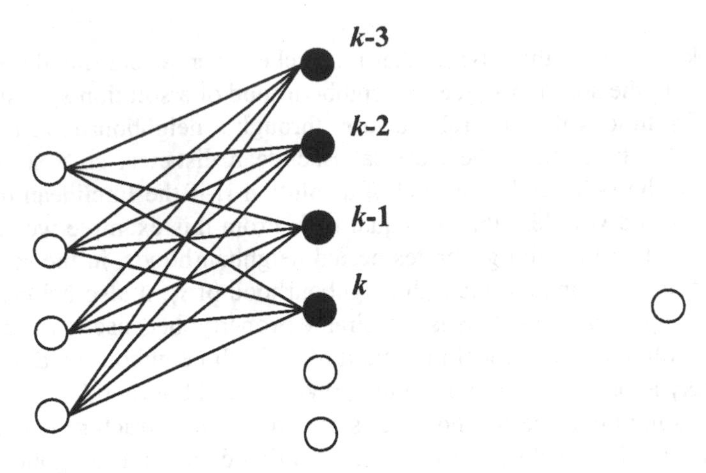

Figure 4-8. Neighbourhood structures for the VNS.

The neighbourhood definition given above satisfies the property of generating nested neighbourhoods, since when increasing the value of k, all the previous neurons, and consequently their incoming weights, are considered. Once a local optimum is reached in any of the neighbourhoods, the partial errors of some weights change and so the neurons ordering. In addition, following the variable neighbourhood scheme, the value of k is set to 1. If the local optimum is good enough, the procedure will eventually consider a neighbourhood containing all the neurons of the net, which is in fact a global search through all the search space.

After deciding the way in which we are going to select a neighbourhood, the next step is to consider the stopping criterion of the search process. Usually, the criterion for stopping the training process for an artificial neural network is the increasing of the error of the test set V_t. However, this condition can be introduced in the "improved" method at the end of each iteration, so a solution would only be accepted if the error of the training set T decreases, and the error of the test set V_t is, at most, the error of the previous best solution for this set. Hence, there is no need for considering it in the stopping criterion, so, in the experimental tests of this chapter, we will use the CPU time. Also, this stopping condition may require more iterations than neurons available in the net, which can lead to inconsistencies in the way neighbourhoods are selected (k can reach a value larger than the neurons in the net, producing an undefined neighbourhood). To solve this, we can slightly modify the neighbourhood function to make it modular, thus selecting the k-th neighbourhood modulus the number of neurons in the net.

The VNS relies on the local search procedure to find a local minimum in a certain neighbourhood. The main advantage of the VNS is that it greatly reduces the search space for the local search procedure. Local search procedures may use either no gradient information at all (as, for example, the "*Flexible Simplex*" method, Nelder and Mead, 1965), first order derivative information (gradient descent, conjugate gradients) or second order derivatives (Newton methods, Levenberg-Marquardt method, etc). The key concept is that gradient information is fast to calculate with the backpropagation technique, and that it is independent of the method used to search the weights-space.

4. EXPERIMENTAL TESTING

In order to test the performance of the VNS for training neural networks, we trained a network to approximate different functions. We compared the results given by a variant of the VNS (the BVNS) and the classical backpropagation, modified to perform a multistart strategy whenever a local miminum is obtained. The problems used consist of several functions, each with different grades of complexity and multiple local minima. These functions are listed below:

$$sexton1: f(x) = x_1 + x_2$$

$$sexton2: f(x) = x_1 * x_2$$

$$sexton3: f(x) = \frac{x_1}{|x_2| + 1}$$

$$sexton4: f(x) = x_1^2 - x_2^3$$

$$sexton5: f(x) = x_1^3 - x_2^2$$

$$Branin: f(x) = \left(x_2 - \left(\frac{5}{4\pi^2} \right) x_1^2 + \left(\frac{5}{\pi} \right) x_1 - 6 \right)^2 + 10 \left(1 - \frac{1}{8\pi} \right) \cos(x_1) + 10$$

$$B2: f(x) = x_1^2 + 2x_2^2 - 0.3\cos(3\pi x_1) - 0.4\cos(4\pi x_2) + 0.7$$

$$Easom: f(x) = -\cos(x_1)\cos(x_2)\exp\left(-\left((x_1 - \pi)^2 + (x_2 - \pi)^2\right)\right)$$

$$Shubert: f(x) = \left(\sum_{j=1}^{5} j\cos((j+1)x_1 + j) \right) \left(\sum_{j=1}^{5} j\cos((j+1)x_2 + j) \right)$$

$$Booth: f(x) = (x_1 + 2x_2 - 7)^2 + (2x_1 + x_2 - 5)^2$$

$$Matyas: f(x) = 0.26(x_1^2 + x_2^2) - 0.48x_1x_2$$

$$Schwefel: f(x) = 418.9829n + \sum_{i=1}^{n} \left(-x_i \sin\sqrt{|x_i|} \right)$$

It is widely known that normalizing both the inputs and the outputs to have zero mean and unit standard deviation improves the training process. Therefore, we normalize the data for the training and test phases using Eq. (2), but we undo the normalization to carry out the validation phase and calculate the errors. The rationale behind this fact is to obtain the final error using the same units as the original data.

$$x' = \frac{x - \bar{x}}{\sigma} \tag{2}$$

An additional consideration has to be taken into account. The normalization of the training set T is done by means of calculating the mean and the standard deviation of each of the input and output variables through all the patterns in the pattern set. However, the net modifies its weights to match the output values of the patterns considering a specific mean and standard deviation vector. So, in order to obtain comparable results both for the test and validation sets V_t, V, they must be rescaled with those same mean and standard deviation vectors used for the training set T.

The training, test and validation sets consist of 200 observations with data randomly drawn from [-100, 100] for x_1 and [-10, 10] for x_2. The architecture of the net consists of two neurons in the input layer (plus the bias), nine neurons in the hidden layer (plus the bias) and one neuron for the output layer.

In order to get local minima after setting the weights, we have used as local optimizers the Flexible Simplex and the GRG (described in Smith and Lasdon, 1992), although only the results of the GRG are shown, due to the higher quality of the solutions obtained.

The stopping criterion was the CPU time, which was set to 10 minutes for statistical purposes, in order to obtain low variance in the results. However, it is interesting to mention that, in most of the cases, a 40 seconds run was enough to obtain a solution which error was roughly 2 units above the error obtained by the 10 minutes run average solution.

For our computational testing, all the programs were implemented in C and run on a Pentium 4 at 2.4 Ghz.

Table 4-1 summarizes the results of the validation set both for the backpropagation and variable neighbourhood search. First column shows the function name. Second and third columns report the mean squared validation errors over ten runs obtained by the BP and VNS, respectively.

Table 4-1. Computational results

Problem	BP	VNS
Sexton 1	1.77 ± 1.60	0.00 ± 0.00
Sexton 2	8.59 ± 3.94	0.05 ± 0.01
Sexton 3	1.68 ± 0.22	3.69 ±0.18
Sexton 4	43.89 ± 9.86	0.17 ± 0.06
Sexton 5	14.30 ± 5.50	0.05 ±0.02
Branin	15.32 ± 0.87	5.36 ± 1.15
B2	21.56 ± 11.97	0.41 ± 0.02
Easom	0.19 ± 0.06	0.00 ± 0.00
Shubert	13.67 ± 0.02	24.95 ± 0.37
Booth	111.17 ± 6.93	0.05 ± 0.02
Matyas	5.25 ± 2.56	0.02 ± 0.01
Schwefel	2.93 ± 0.53	0.97 ± 0.04

5. CONCLUSIONS

In this chapter we have described the implementation of a variable neighbourhood search for training a feed-forward neural network with one hidden layer.

The results show that a significant improvement on the quality of the solutions is obtained when using the VNS with the GRG as local optimizer. The existence of several variants of the VNS gives a wide set of possible implementations, each of them with its own properties (e. g., VND for faster, lower quality solutions; GVNS for slower, higher quality solutions). However, and independently of the variant used, the definition of the neighbourhood structures provided in this chapter allows for a better understanding of the optimization process (related to neurons, instead of isolated weights), reducing at the same time the average search space size in an efficient way.

ACKNOWLEDGEMENTS

The authors want to acknowledge Rafael Martí and Abdellah El Fallahi for their support and for supplying the GRG code for the computational tests.

This research has been partially supported by the projects TIC2002-04242-C03-01, TIN2005-24056-E (70% of which are FEDER funds) and TIN2004-20061-E (Spanish Heuristic Network).

REFERENCES

Blum, E. K., and Li, L. K., 1991, Approximation theory and feedforward networks, *Neural Networks* **4**(4):511-515.

Brimberg, J., Hansen, P., Mladenović, N., and Taillard, E., 2000, Improvements and comparison of heuristics for solving the multisource Weber problem, *Operations Research* **48**:444-460.

El–Fallahi, A., Martí, R., and Lasdon, L., 2005, Path relinking and GRG for artificial neural networks, *European Journal of Operational Research* (to appear).

Gallant, A. R., and White, H., 1992, On learning the derivatives of an unknown mapping with multilayer feedforward networks, *Neural Networks* **5**:129-138.

Glover, F., and Kochenberger, G., 2003, *Handbook of Metaheuristics*, Kluwer.

Hansen, P., and Mladenović, N., 2001, Variable neighbourhood search: principles and applications, *European Journal of Operational Research* **130**:449-467.

Hansen, P., and Mladenović, N., 2003, Variable neighbourhood search, in: *Handbook of Metaheuristics,* F. Glover, and G. Kochenberger, eds., Kluwer, pp. 145-184.

Johnson, D. S., and McGeoch, L. A., 1997, The travelling salesman problem: A case study in local optimization, in: *Local Search in Combinatorial Optimization*, E. H. L. Aarts, and J. K. Lenstra, eds., John Wiley & Sons, pp. 215-310.

Lourenco, H. R., Martin, O., and Stuetzle, T., 2003, Iterated Local Search, in: *Handbook of Metaheuristics,* Glover, F., and Kochenberger, G., eds., Kluwer, pp. 321-353.

Mladenović, N., and Hansen, P., 1997, Variable neighbourhood search, *Computers and Operations Research* **24**:1097-1100.

Nelder, J. A., and Mead, R., 1965, A Simplex method for function minimization, *Computer Journal* **7**:308-313.

Reeves, C. R., 1993, *Modern Heuristic Techniques for Combinatorial Problems*, Blackwell Scientific Press.

Rumelhart, D. E., Hilton, G. E., and Williams, R. J., 1986, Learning internal representations by error propagation, in: D. E. Rumhelhart, J. L. McClelland, and the PDP Research Group, *Parallel Distributed Processing: Explorations in the Microstructure of Cognition, Volume 1: Foundations*, Cambridge, MA: MIT Press. Reprinted in Aderson and Rosenfeld (1988), pp. 318-362.

Smith, S., and Lasdon L., 1992, Solving large nonlinear programs using GRG, *ORSA Journal on Computing* **4**(1):2 - 15.

Chapter 5

ESTIMATION OF DISTRIBUTION ALGORITHMS

Julio Madera[1] and Bernabé Dorronsoro[2]

[1]*Department of Computing, University of Camagüey, Cuba;*

[2]*Department of Computer Science, University of Málaga, Spain*

Abstract: Training Artificial Neural Networks (ANNs) is a very complex task with a high practical relevance in the field of supervised learning. In this chapter, the problem of training ANNs is faced with several Estimation of Distribution Algorithms (EDAs) with different features, exploring both continuous and discrete search spaces. These EDAs have been tested on a benchmark taken from the medicine field. The results have been carefully analyzed, and compared versus those of other algorithms in the literature for the considered problem. Our conclusions are both that our EDAs are competitive with the other compared algorithms, and also that the use of continuous EDAs is advantageous, in general, versus discretizing the search space for the studied problems.

Keywords: Discrete EDAs; continuous EDAs; supervised learning.

1. INTRODUCTION

Artificial Neural Networks (ANNs) are adaptive learning tools particularly good at prediction tasks, and having high computational abilities. Thus, the importance of Artificial Neural Networks (ANNs) for the research community has grown during the last decade. They have been applied in many fields of experimental science, such as physics (Dawn, 1994), botany (Simpson, 1992), or medicine (Fujita et al., 1992). However, the main handicap of ANNs is their inherent dependency on the training sets. In the training process, they learn on their own (by adjusting a set of weights) the implicit relationship between input and output data, which have to be known in advance. This training task constitutes an essential process in supervised learning.

In the literature, many different approaches have been proposed for solving the problem of training ANNs. Among them, one of the most popular one is the *back-propagation* method (Rumelhart, 1986), which tunes the values of the set of weights by following a gradient descent method. More recently, other modern techniques are being used for training ANNs. Specifically, Evolutionary Algorithms (EAs) have been successfully applied to this problem (Cotta et al., 2002; Cantú-Paz, 2003; Alba and Chicano, 2004). These new stochastic techniques perform an *intelligent* search, obtaining very accurate solutions much more quickly than other classical methods. This allows us to face harder problems with respect to other deterministic algorithms.

This chapter is mainly motivated by the use of Estimation of Distribution Algorithms (EDAs), a kind of EA, to face the problem of training ANNs. This problem has not been deeply investigated with this kind of algorithms, and only a few interesting works facing the problem with EDAs can be found in the literature. In the work of Baluja (1995) several optimization techniques are applied to this problem, including the PBIL algorithm (a kind of EDA), among others. Additionally, the work of Zhang and Cho (2000) proposes the use of evolutionary Bayesian algorithms to adjust the weights of a neural network (applied to time series predictions) with a tree structure. The PBIL algorithm is also used by Galic and Höhfeld (1996) and Maxwell and Anderson (1999) for solving the training problem of different ANNs (a multilayer perceptron network, and a hidden Markov model, respectively). Finally, in the work of Cotta et al. (2002) two EDAs, $UMDA_C$ and $MIMIC_C$, are applied to this problem, obtaining competitive results with those of other classical evolutionary algorithms.

The contribution of this work is twofold. Firstly, we solve the problem with six different EDAs in both the discrete and continuous domains, analyzing the behavior of all of them. Secondly, we will compare the results of our EDAs with those of other algorithms in the literature for the same problem instances.

The chapter is organized as follows. In section 2 the Artificial Neural Network computational model is introduced. After that, section 3 describes the different EDAs we analyze. Our experimental results are shown in section 4 and, finally, in section 5 conclusions and future work are given.

2. ARTIFICIAL NEURAL NETWORKS

Artificial Neural Networks are inspired in the bioelectrical networks in the brain formed by neurons and their synapses. In an ANN model, simple nodes (called *artificial neurons*) are connected together to form a network of nodes. ANNs use a mathematical or computational model for performing a parallel processing of the information (McClelland and Rumelhart, 1996), and the global behavior of an ANN is determined by the connections between the processing elements (nodes) and their parameters (e.g., the weights of the connections between neurons).

Artificial neurons compute an output from their inputs by using an activation function (see Fig. 5-1). The input of a neuron is composed of both the weighted output of other neurons (or input of the ANN for the neurons in the input layer), and a bias term (only one per neuron). A bias is similar in function to a threshold and is treated as a weight connected to a node that is always on. The output of a neuron will be either processed by other neurons or part of the ANN response (if the neuron belongs to the output layer).

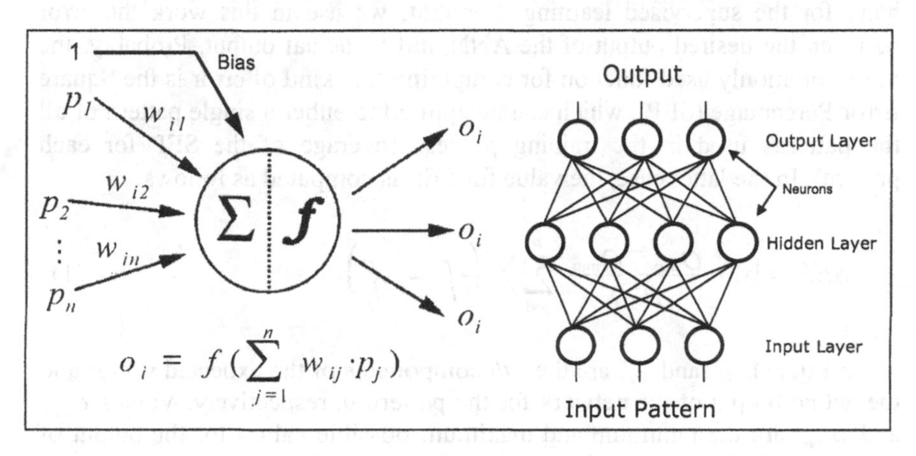

Figure 5-1. Examples of an artificial neuron (left) and a general multilayer perceptron (right).

When designing a neural network, there are several important decisions to take. For example, both the architecture of the network and the weights of the connections between artificial neurons have to be decided, among others.

There are typically two main kinds of network architectures. One of them is *feedforward* networks, in which neurons never send information to others from which the former receives an input (connections are strictly feedforward). The other kind of network architecture is *recurrent* networks, in which feedback connections are allowed between neurons. In this work, we focus on feedforward networks. Specifically, we will concentrate on the Multilayer Perceptron Model (MLP) (Rosenblatt, 1962), in which artificial neurons are ordered in layers and connections are only allowed between adjacent layers, as shown in Fig. 5-1. Every artificial neuron of a layer is connected to all the neurons of the adjacent layer with a given weight w_{ij}.

When designing an MLP network, several parameters have to be defined, as the number of layers and the number of neurons per layer. After designing the structure of our MLP network, the training process has to be carried out for adjusting simultaneously both the weights of the connections between neurons and the bias terms. The values of these parameters to adjust will determine the behavior of the network, so we have to take special care on the process of tuning them, which is the case of study in this work. This training task is performed by adjusting the weights of the network such that for a given set of input patterns the network returns a desired set of outputs. Both the input and output patterns have to be known beforehands, and they compose the so called training set.

There is a need of establishing a measure for the quality of the network at hand for the supervised learning. For that, we use in this work the error between the desired output of the ANN and its actual output. Probably, the most commonly used function for computing this kind of error is the Square Error Percentage (SEP), which can be applied to either a single pattern or all the patterns used in the training process (average of the SEP for each pattern). In the latter case, the value for SEP is computed as follows:

$$SEP = 100 \cdot \frac{o_{max} - o_{min}}{P \cdot S} \cdot \sum_{p=1}^{P} \sum_{i=1}^{S} \left(t_i^p - o_i^p \right)^2 \quad (1)$$

In Eq. (1), t_i^p and o_i^p are the *i-th* components of the expected vector and the actual output of the network for the pattern p, respectively. Values o_{min} and o_{max} are the minimum and maximum possible values for the output of every neuron, P stands for the number of patterns composing the training set, and S is the number of neurons composing the output layer.

Besides the error between the desired and the actual outputs, it is also interesting in classification problems to measure the quality of the network in terms of the patterns that have been incorrectly classified after the training process. That measure is called the Classified Error Percentage (CEP), and it is used in this work for measuring the quality of the final ANNs. Essentially, CEP is a measure of the quality of the training process for the ANN. Thus, this value will allow us to compare the results of the studied algorithms.

3. ESTIMATION OF DISTRIBUTION ALGORITHMS

Evolutionary Algorithms -EAs- (Bäck et al., 1997) are stochastic search techniques designed as an attempt to solve adaptive and hard optimization tasks on computers. In fact, it is possible to find this kind of algorithms applied for solving complex problems like constrained optimization tasks, problems with a noisy objective function, or problems having high epistasis and multimodality (Alba and Dorronsoro, 2005). These algorithms work over a set (*population*) of potential solutions (*individuals*) by applying some stochastic operators, called *variation operators* (e.g., natural selection, recombination, or mutation), on them in order to search for the best solutions.

Estimation of Distribution Algorithms (EDAs) are an alternative family to traditional EAs in which no variation operators are used. The EDA class of algorithms was introduced in the work of (Müehlenbein and Paaβ, 1996) as an extension to genetic algorithms (GAs). Their motivation was influenced by two principal questions. Firstly, the difficulty that GAs find when facing problems with deception and not separability. Secondly, because the implicit search carried out by the variation operators in GAs can be extended by including information about the correlation of the variables in the set of individuals with high fitness values. Therefore, the successive generations of individuals are created by using estimations of distributions observed in the current population instead of evolving the population with the variation operators (Larrañaga and Lozano, 2002; Mühlenbein et al., 1999), like in the case of other EAs. This new class of evolutionary algorithms is also known in the literature as Probabilistic Model Building Genetic Algorithms -PBMGA- (Pelikan et al., 1999b) and Iterated Density Estimators Evolutionary Algorithms -IDEAS- (Bosman and Thierens, 1999).

In Table 5-1 we show a brief pseudocode for an EDA. As it can be seen, an EDA starts by randomly generating the initial population of potential solutions (*individuals*, also called *points* in the literature of EDAs), and then the algorithm begins to iteratively evolve the current population until a termination condition is met. This termination condition is usually either to find a solution or to reach a maximum number of function evaluations. The way in which the population evolves is, precisely, the main difference between the family of EDAs and other traditional kinds of EAs. The new generation of individuals is computed as follows. From the whole population, only M individuals (with $M \leq N$, being N the size of the population) are selected. Then, the EDA explicitly extracts global statistical information from this set of M parent solutions, and build a posterior probability distribution model of promising solutions $p^s(x,t)$, based on the extracted information (s, x, and t represent the selected set of parents, the set of variables composing the individuals, and the generation number, respectively). After estimating the probability distribution, N new solutions are sampled from the model thus built and fully or in part replace the current population (at generation t) to form the new one (generation $t+1$).

Table 5-1. Pseudocode for an EDA algorithm

```
EDA
Set t ← 1;
Generate N >> 0 individuals randomly;
while termination condition is not met do
    Select M ≤ N individuals according to a selection method;
    Estimate de distribution p^s(x,t) of the selected M individuals;
    Generate N new individuals according to the distribution p^s(x,t);
    Set t ← t+1;
end while
```

The main steps characterizing EDAs are both to estimate the distribution model of the selected solutions, $p^S(x,t)$, and to generate new points according to this distribution. The behavior of an EDA (and thus the obtained results) is highly influenced by the way in which the mentioned probability distribution is estimated. For that, graphical models have became common tools capable of efficiently representing the probability distribution. Usually, EDAs work with fixed-length strings of discrete or continuous variables, so different techniques have been proposed for representing the probability distribution depending on whether the domain is discretized or continuous. When the variables are discretized, some authors (Pelikan et al., 1999a; Soto et al., 1999; Müehlenbein and Mahnig, 2000) have proposed the use of Bayesian networks of different complexities to represent the probability distribution. In these Bayesian networks, nodes are the variables and the edges represent the relations between variables. Conversely, Gaussian networks are usually employed for continuous domains (Larrañaga et al., 1999).

Due to the capital importance of the estimation of the distribution model in the behavior of EDAs and the high cost of this operation, some different improvements have been proposed for this issue in the literature of EDAs. Many algorithms use a score+search procedure for learning the probability distribution. These methods define a metric that measures the goodness of every candidate Bayesian network with respect to a database of cases. In this case, a search procedure to move efficiently through the space of possible networks is needed. Additionally, some other algorithms perform independence tests on the variables of the problem in order to take advantage of the dependencies and independencies found by inserting or deleting edges to the Bayesian network, respectively.

The application of EDAs to optimization problems has been intense during the last decade. This has been motivated in part by the fact that in many of the reported results, EDAs overcome other evolutionary approaches such as genetic algorithms (Larrañaga and Lozano, 2002; Pelikan and Mülenbein, 1999; Cantú-Paz, 2002). However, EDAs usually have high computational costs compared to other more classical evolutionary algorithms. Thus, due to these high computational costs of many EDAs, the current state-of-art of the field requires the development of new and more powerful strategies for implementing them.

From the theoretical point of view, Mühlenbein and Höns (2005) recently showed evidences of the relation between EDAs and other research fields such as the statistics, belief networks and statistical physics. These disciplines can help to understand much better the dynamics of EDAs, and hence, to develop new powerful algorithms. At the same time they distinguish two branches in the design of EDAs. In the first branch the factorization of the distribution is computed from the structure of the function to be optimized. In the second one, the structure is computed from the correlations of the data.

Now, we are in condition to make a classification of EDAs according to the type of interaction between allele values (variables) allowed in the model of the probability distribution (Larrañaga and Lozano, 2002). Thus, we distinguish among EDAs without dependences between variables, with pairwise interactions, or with multiple interactions:

1. **Without dependencies:** this is the simplest way to estimate the distribution of promising solutions. In this kind of EDA, it is assumed that all the problem variables are independent one each other. Mathematically, the estimation of distribution for EDAs without dependences is $p^s(x,t) = \prod_{i=1}^{n}(x_i,t)$. These algorithms with no dependences are appropriate for solving linear problems, where the variables are not mutually interacting, but they fail on problems with strong dependences among variables because they do not take into account them. Some examples of existing algorithms of this kind are the Univariate Marginal Distribution Algorithm, UMDA (Mühlenbein and Voigt, 1996; Mühlenbein, 1997), or the Population Based Incremental Learning algorithm, PBIL (Baluja, 1994).
2. **Bivariate dependencies:** this kind of EDA was the first one assuming that the variables in a problem are not independent. These algorithms compute the estimation of distribution taking into consideration existing dependencies between pairs of variables. Bivariate EDAs reproduce and mix building blocks of order two very efficiently; therefore, they are specially suitable for solving linear and quadratic problems. From this kind of algorithms, we will highlight the Mutual Information Maximization Algorithm for Input Clustering, MIMIC (De Bonet et al., 1997), MIMIC_C, an adaptation of MIMIC for continuous domain (Larrañaga et al., 1999, 2000) and the Bivariate Marginal Distribution Algorithm, BMDA (Pelikan and Mühlenbein, 1999).

3. **Multiple dependencies:** bivariate dependencies are usually insufficient for solving problems with multivariate of highly-overlapping building blocks, so research in this area continued with more complex models. In the multiple dependencies case, all of the possible dependencies between variables are taken into consideration, with no restrictions in the complexity required by the learning algorithm. These algorithms achieve a good performance on a wide range of decomposable problems, although they can still find difficulties when the problems are decomposable into terms of bounded order. Examples for this class of EDAs are the Bayesian Optimization Algorithm, BOA (Pelikan et al., 1999a, 2000a, 2000b; Pelikan and Goldberg, 2000), the Polytree Approximation of Distribution Algorithm, PADA (Soto et al., 1999), and the Estimation of Gaussian Networks Algorithm, $EGNA_{BGe}$ (Larrañaga et al., 1999, 2000).

We graphically show in Fig. 5-2 the relations between variables taken into account by different EDAs for modeling the probability distribution without dependences (UMDA), with bivariate dependences (MIMIC), and with multiple dependencies (BOA). Additionally, there exist other different structures, like the case of ECGA, in which the variables are divided into a number of mutually independent clusters, being each one taken as a whole.

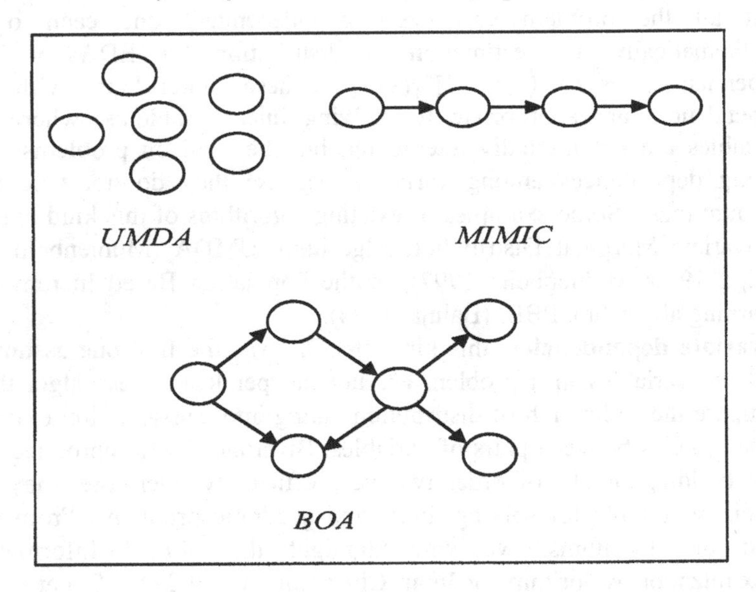

Figure 5-2. Relations between variables for some different EDAs.

In the remaining of this section we briefly explain the six EDAs we have used in this work for solving the problem of training ANNs. Three of them explore a discretized search space (section 3.1), while the other three ones work in the continuous domain (section 3.2). In the two cases, we propose EDAs considering univariate, bivariate and multivariate dependencies.

3.1 EDAs in Discrete Domains

In this section, we present the three EDAs with discretized variables we study in this work. The three algorithms differ in the complexity of the model they use to learn the probability distribution for the selected individuals. They are UMDA, having no dependencies among variables, MIMIC, which use bivariate dependencies, and BOA, an EDA with multiple dependencies.

3.1.1 UMDA Algorithm

The Univariate Marginal Distribution Algorithm (UMDA) (Mühlenbein and Voigt, 1996) is one of the simplest algorithm in the EDAs family. In UMDA, it is considered that variables are independent from the others, so there is not any dependency between them. The current generation evolves towards the new one by computing the frequencies of values of the variables on each position in the selected set of promising solutions. These frequencies are then used to compute new solutions, which replace the old ones. Due to its simplicity, it is a very efficient algorithm (converges quickly), and its behavior is particularly good for linear problems.

3.1.2 MIMIC Algorithm

The Mutual Information Maximization Algorithm for Input Clustering (MIMIC) was proposed by De Bonet et al. (1997). This algorithm considers pairwise interactions between variables for computing the probability of distribution. MIMIC uses a simple chain distribution (a permutation of the variables) that maximizes the so-called mutual information of neighboring variables (string positions). Hence, the distance between the chain and the complete joint distribution has to be minimized, and the Kullback-Leiber divergence is used for measuring this distance. The optimization task is solved with a greedy algorithm, which is a very efficient tool for this purpose (avoids an exhaustive search), although the global optimality of the distribution is not guaranteed.

3.1.3 BOA Algorithm

Finally, the multivariate EDA for discrete domains we study in this work is the Bayesian Optimization Algorithm (BOA), proposed by Pelikan et al. (1999a, 2000a, 2000b) and Pelikan and Goldberg (2000). In this algorithm, the distribution probability is estimated by means of Bayesian networks, which are measured with a Bayesian metric (Bayesian Dirichlet equivalence) in order to compute how adequate the found structures are. In each generation, the network is built using a greedy algorithm which begins from scratch. There exists a parameter (k) to limit the search through the space of all possible structures. The value of this parameter means the maximum number of parents for any node of the network.

3.2 EDAs in Continuous Domains

In contrast to section 3.1, the algorithms proposed here use variables belonging to the continuous domain, so the search space is not discretized in this case. Like in the case of section 3.1, we propose three EDAs differing in the dependencies taken into account among variables.

3.2.1 UMDA$_C$ Algorithm

The UMDA$_c$ algorithm, proposed by Larrañaga et al. (1999, 2000), is a modified version of UMDA for continuous domain. Hence, the algorithm assumes that the variables are independent one each other, and some statistical tests are carried out for every variable in each generation in order to find the density function that better fits this variable (some different density functions are considered). The two parameters that are estimated in case that all the distributions are normal are the average $\mu_i(t)$ and the standard deviation $\sigma_i(t)$.

3.2.2 MIMIC$_C$ Algorithm

MIMIC$_c$ is an adaptation of the MIMIC algorithm applied to continuous domains. It was proposed for the first time by Larrañaga et al. (1999, 2000), and the idea is to estimate the distribution of each of the variables as closely as possible taking into account dependencies between pairs of variables. Hence, the probability of distribution is assumed to be a bivariate Gasussian.

3.2.3 EGNA$_{BGe}$ Algorithm

The Estimation of Gaussian Networks Algorithm (EGNA$_{BGe}$), proposed by Larrañaga et al. (1999, 2000), is an EDA with multivariate dependencies among variables for continuous optimization. The model induction is carried out through a local search method plus the Bayesian Gaussian Equivalence (BGe) scoring metric (Geiger and Heckerman, 1994). The BGe metric consists in computing a Bayesian score for using later the same values for Gaussian networks reflecting identical conditional (in)dependencies. The local search is used for looking for the most appropriate Gaussian network.

4. EXPERIMENTAL STUDY

In this section, we analyze the experimental results we have obtained in our work when solving the supervised training of ANNs with the six different proposed EDAs. In section 4.1, we begin by introducing the benchmark we have selected for the experiments of this study, as well as the parameterization used for the algorithms in our tests. Later, section 4.2 contains the results of all our experiments, which are carefully analyzed. Finally, we compare in section 4.3 our results with those obtained by other authors with different algorithms for the same problem instances.

4.1 Computational Experiments

The benchmark we have selected for this study belongs to the field of medicine, and it is composed of four different classification problems: diagnosis of breast cancer, diagnosis of diabetes in Pima Indians, diagnosis of heart disease, and diagnosis of thyroid hyper- or hypofunction. In classification problems, the objective is to determine the class that an input vector belongs to (Prechelt, 1994). For that, both the weights of the network and the bias terms have to be tuned such that for each input pattern of the training set the network returns its corresponding output (the training set is a collection of inputs and their desired outputs). There exist several ways for interpreting the class that an input belongs to in ANNs. In this work we will assign an output neuron for each class, and the method for interpreting the output of the network is the so called *winner-takes-all*, which consists in that the output neuron with the largest value determines the class.

The studied problems are selected from a well-known benchmark based on real data, called PROBEN1 (Prechelt, 1994). Hereinafter, a short description of each problem is given:

- **Cancer:** Diagnosis of breast cancer. It consists in classifying a tumor as either benign or malignant, based on cell descriptions gathered by microscopic examination. The data base is composed of 699 examples, originally obtained from the *Univ.Wisconsin Hospitals, Madison*, from Dr. William H. Wolberg (Bennett and Mangasarian, 1992; Mangasarian et al., 1990; Wolberg, 1990; Wolberg and Mangasarian, 1990).
- **Diabetes:** Diagnosis of Pima Indians diabetes. Based on medical examinations, it has to be decided whether a Pima Indian individual is diabetes positive or not. The data base was originally obtained from the *National Institute of Diabetes and Digestive and Kidney Diseases* by V. Sigillito (Smith et al., 1988), and it is composed of 768 examples.
- **Heart:** Prediction of heart disease. It consists in deciding whether at least one of four major vessels is reduced in diameter by more than 50%. This decision is made based on personal data and results of medical examinations. There are 920 examples from four different sources: *Hungarian Institute of Cardiology* in Budapest (Andras Janosi, M.D.), *University Hospital of Zurich* in Switzerland (William Steinbrunn, M.D.), *University Hospital of Basel* in Switzerland (Mathhias Pfisterer, M.D.), and the *V.A. Medical Center of Long Beach and Cleveland Clinic Foundation* (Detrano et al., 1989; Gennari et al., 1989).
- **Thyroid:** This last problem deals with diagnosing the thyroid function of a patient as being overfunction, normal function, or underfunction. Hence, in this problem the input pattern has to be classified into one of three possible classes, in contrast to the other three proposed problems, for which there are only two output classes. Additionally, this dataset is also characterized by the large set of patterns used with respect to the other three studied problems (see Table 5-2).

The architecture of the MLPs used for the four proposed problems are composed of three layers: the input and output ones, and an additional hidden layer. In all the problems, the number of artificial neurons composing this hidden layer is set to six, while in the case of the input and output layers it depends on the problem instance. The number of neurons of the input layer is 9 for Cancer, 8 for Diabetes, 35 for Heart, and 21 for Thyroid, while the output layer is composed of two neurons for Cancer, Diabetes and Heart, and three in the case of Thyroid. The activation function used in every artificial neuron of the network is the sigmoid function.

The pattern set is split into two sub-groups: one for training the ANN, and another for testing the behavior of the trained network. The training set approximately suppose 75% of all patterns in the database, while the testing set is the remaining 25% of the patterns. We show in Table 5-2 the architecture of the network (number of neurons at the input-hidden-output layer), and the number of training and test patterns for each instance.

Table 5-2. MLP structure and pattern composition for all instances

Instance	Architecture of ANN	Patterns	
		Training	Test
Cancer	9 - 6 - 2	525	174
Diabetes	8 - 6 - 2	576	192
Heart	35 - 6 - 2	690	230
Thyroid	21 - 6 - 3	5400	1800

Finally, in the remaining of this section we will present the parameters used in our EDAs for facing the problems. Some of these parameters are common for all the studied EDAs (e.g., population size, truncation threshold, or the maximum number of evaluations), while others are specific to the algorithms. In the case of the common ones, we have set the same values for them in all the algorithms (detailed in Table 5-3). The exception is BOA, for which we have used some specific parameters in order to reduce the high quantity of memory consumed due to the complex model it builds. Several preliminary experiments were performed for tuning all the parameters, adopting the values which offered the best results.

The selection method used in all the studied EDAs is the truncation selection model, which consists in selecting the best $\tau \cdot 100\%$ individuals of the current population. The parameter τ is called the truncation threshold, and it has been set to 0.5 for our experiments. The population contains 64 individuals (1024 in BOA), and the best one is maintained in the population for successive generations (elitism). The algorithms stop when a maximum of 5000 evaluations are performed (52174 in the case of BOA).

Table 5-3. Common parameters for the EDAs algorithms

Parameter	Cancer	Diabetes	Heart	Thyroid
Population size	64 individuals (1024 for BOA)			
Selection method	Truncation selection model			
Truncation threshold	0.5 (the 50% of best individuals are chosen)			
Elitism	Best individual survives in next generation			
Fitness evaluations	5000 fitness function evaluations (52174 for BOA)			

In the case of the EDAs in discrete domains, we have been forced to discretize the values for the weights of the ANN. Hence, we have coded these weights with 16-bit (12-bit in BOA) binary vectors. Each weight is a real number defined into the interval [-1.0,1.0]. Consequently, although we model the Cancer, Diabetes, Heart, and Thyroid problems with 74, 68, 230, and 152 variables, respectively, in the continuous case, we need 1184, 1088, 3680, and 2432 bits for the same problems when using a discretized search space (888, 816, 2760, and 1824 bits in the case of BOA). Thus, the models that the EDAs construct for estimating the distribution of the variables are much more complex in the discrete case. In the case of BOA, a maximum of $k = 3$ parents is allowed for each variable in the Bayesian network.

The fitness function we want to maximize for solving the problem is the negative value of SEP (SEP is defined in Eq. 1).

4.2 Analysis of the Results

In this section we present the results obtained in all our experiments. The figures in tables 5-4 and 5-5 stand for the average ($\overline{X}$) and standard deviation (σ_n) of the CEP value (percentage of wrongly classified patterns) obtained by the six EDAs (both in discrete and continuous domains) for the problems in our benchmark. Additionally, we show in Table 5-6 the obtained average execution times. All these values are computed after making 50 independent runs by every algorithm for each problem, and ANOVA tests have been performed in order to obtain statistically significant values in our tests, at a confidence level of 95%. Symbol '+' in tables 5-4 and 5-5 stand for statistically significant differences (p-values under 0.05), while '-' means no significant differences.

As it can be seen in Table 5-4, UMDA obtains the best results for the easiest problem (Cancer), and BOA finds also the best solutions for Diabetes and Thyroid. These three results are statistically significant. In the case of Heart, MIMIC is the algorithm throwing the best results, but there are no significant differences. It stands out the different behavior of the studied algorithms for the proposed problems. We can say generally speaking that the algorithms considering higher degrees of dependencies among variables are more suitable for solving the most complex problems, since they converge more slowly, thus maintaining the diversity for longer. Probably, this slow convergence is also the cause for the bad results (compared to the other studied EDAs) these algorithms obtain for the simplest problems.

Table 5-4. EDAs in discrete domain. Percentage of wrongly classified patterns (CEP)

CEP (%)		*UMDA*	*MIMIC*	*BOA*	*ANOVA*
Cancer	$\overline{X}_{\pm\sigma_n}$	$1.54_{\pm 0.57}$	$1.69_{\pm 0.94}$	$3.07_{\pm 2.34}$	+
Diabetes	$\overline{X}_{\pm\sigma_n}$	$47.70_{\pm 5.01}$	$46.80_{\pm 2.48}$	$44.99_{\pm 2.42}$	+
Heart	$\overline{X}_{\pm\sigma_n}$	$38.83_{\pm 3.77}$	$37.69_{\pm 2.78}$	$38.47_{\pm 3.77}$	-
Thyroid	$\overline{X}_{\pm\sigma_n}$	$8.53_{\pm 0.95}$	$8.22_{\pm 0.83}$	$7.28_{\pm 0.02}$	+

Table 5-5. EDAs in continuous domain. Percentage of wrongly classified patterns (CEP)

CEP (%)		*UMDA$_C$*	*MIMIC$_C$*	*EGNA$_{BGe}$*	*ANOVA*
Cancer	$\overline{X}_{\pm\sigma_n}$	$2.54_{\pm 4.27}$	$1.36_{\pm 0.85}$	$1.61_{\pm 1.10}$	-
Diabetes	$\overline{X}_{\pm\sigma_n}$	$46.36_{\pm 0.97}$	$46.42_{\pm 0.83}$	$46.53_{\pm 1.12}$	-
Heart	$\overline{X}_{\pm\sigma_n}$	$36.34_{\pm 2.00}$	$37.10_{\pm 1.46}$	$36.35_{\pm 1.99}$	+
Thyroid	$\overline{X}_{\pm\sigma_n}$	$7.44_{\pm 0.22}$	$7.73_{\pm 0.35}$	$7.45_{\pm 0.28}$	-

Related to the algorithms working in continuous domains (results are given in Table 5-5), the reader can see that $UMDA_C$ is the best of the studied algorithms for Diabetes, Heart and Thyroid, but we only found statistical advantages in the case of Heart, for which the result of $MIMIC_C$ is (statistically) worse than those of the other two algorithms. Precisely, $MIMIC_C$ has been the best algorithm for the other problem, but once again there are not significant differences.

In Fig. 5-3 we plot the average solutions obtained by all the studied algorithms for the proposed problems. Comparing the six algorithms, we could think that there are not important differences among their results, but after computing the ANOVA test we have obtained statistically significant differences for all the problems. These statistical differences were generally found between the discrete EDAs and their continuous counterparts (as well as among the discrete algorithms themselves, as stated hereinbefore). BOA is the best (significantly) of the six algorithms for Diabetes and Thyroid, but it is the worst one also with statistical significance (except for $UMDA_C$) in the case of the Cancer problem. Regarding the Heart problem, the results obtained by the univariate and multivariate studied algorithms in continuous domains ($UMDA_C$ and $EGNA_{BGe}$) are significantly better than those of their discrete counterparts (UMDA and BOA). With respect to MIMIC and $MIMIC_C$, there are no statistical differences either between them nor with the other algorithms for this problem.

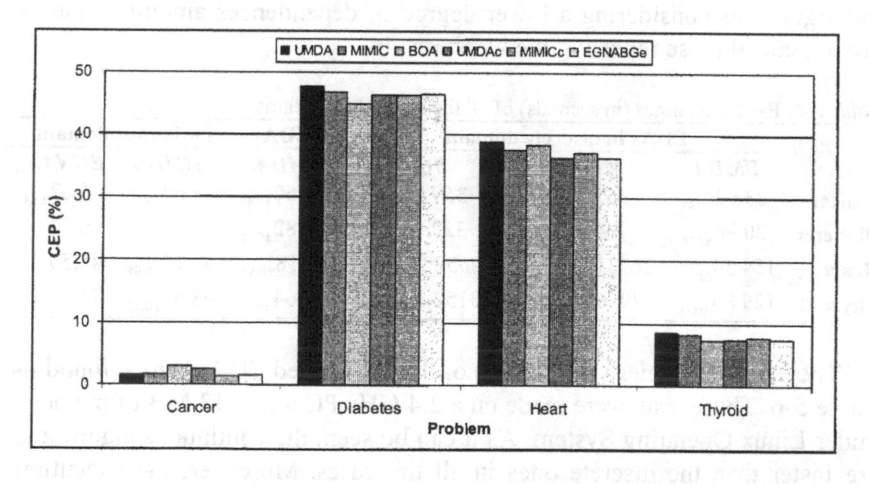

Figure 5-3. CEP values for all the algorithms.

If we analyze the results of the univariate, bivariate and multivariate algorithms in discrete and continuous domains, we obtain (with statistical significance) that $UMDA_C$ is better than UMDA for Heart and Thyroid, $MIMIC_C$ is better than MIMIC for Thyroid, and $EGNA_{BGe}$ is better than BOA for Cancer and Heart, but worse in the cases of Diabetes and Thyroid (although the reader must remember at this point that BOA has a population 16 times larger than $EGNA_{BGe}$, and computes 10 times more fitness evaluations). In the other cases, there are no significant differences between the algorithms. Hence, we can conclude that the use of continuous variables can be advantageous, in general, regarding the average solutions found. Probably, the worst behavior of the discrete algorithms is due to the large set of variables that they have to manage for computing the distribution probabilities, since each continuous variable is discretized with 16-bit strings and, hence 16 binary variables (12 in the case of BOA).

As it can be seen in Fig. 5-3, the Cancer problem, which is one of the smallest studied instances, is the easiest problem to solve for the compared EDAs. The complexity grows with the size of instances, as can be seen in the cases of Thyroid and Heart. The exception is the Diabetes problem, which is the smallest of the studied intances, but also the one for which the EDAs report the highest percentage of wrongly classified patterns, i.e., it is the most difficult problem to solve. The reason for the bad (although competitive in the literature) results in Heart and Diabetes compared to the other two problems is that the EDAs were not able to find a correct interrelation among the weigths. In that sense, one can see in Fig. 5-3 that the algorithms considering a lower degree of dependences among variables are in general those reporting the worst results.

Table 5-6. Execution times (in seconds) of all the studied algorithms

	EDAs in discrete domain			**EDAs in continuous domain**		
	UMDA	***MIMIC***	***BOA***	***UMDA$_C$***	***MIMIC$_C$***	***EGNA$_{BGe}$***
Cancer	$23.02_{\pm 0.47}$	$962.26_{\pm 7.42}$	$359.90_{\pm 5.31}$	$6.60_{\pm 0.49}$	$8.60_{\pm 0.57}$	$6.62_{\pm 0.49}$
Diabetes	$20.98_{\pm 0.14}$	$764.06_{\pm 8.07}$	$322.16_{\pm 3.24}$	$6.82_{\pm 0.38}$	$9.82_{\pm 0.42}$	$6.74_{\pm 0.44}$
Heart	$154.34_{\pm 1.36}$	$26731.02_{\pm 124.18}$	$24652.94_{\pm 224.00}$	$45.18_{\pm 0.52}$	$58.18_{\pm 0.64}$	$45.50_{\pm 0.61}$
Thyroid	$129.80_{\pm 0.53}$	$7966.74_{\pm 30.46}$	$3152.80_{\pm 69.97}$	$76.64_{\pm 0.80}$	$98.64_{\pm 0.75}$	$75.88_{\pm 0.62}$

The execution times (in seconds) of the six studied EDAs can be found in Table 5-6. These tests were made on a 2.4 GHz PC with 512 MB of memory under Linux Operating System. As it can be seen, the continuous algorithms are faster than the discrete ones in all the cases. Moreover, the execution times of MIMIC and BOA for instances Heart and Thyroid are excessively large, while those problems are solved in reasonable times by the EDAs in continuous domains.

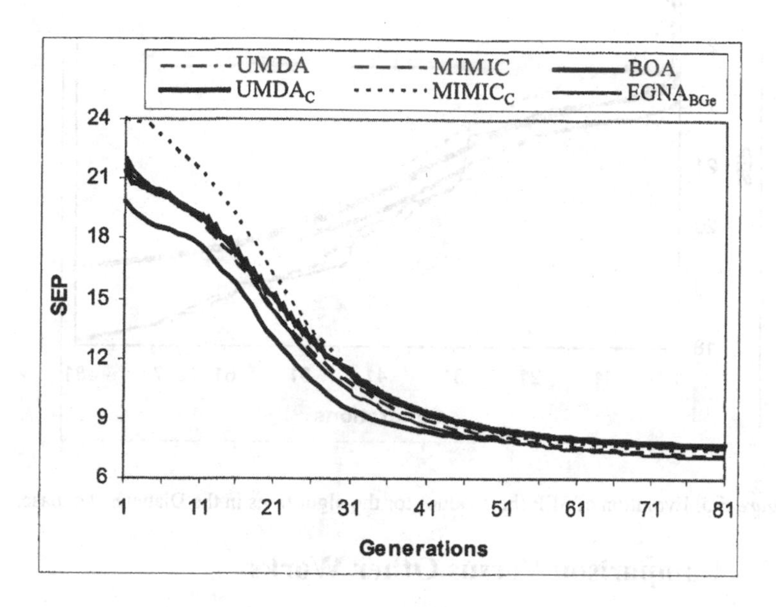

Figure 5-4. Evolution of SEP (best value) for the proposed algorithms in the Cancer database.

Finally, we are also interested in the evolution of the value of the fitness function (SEP) during the execution for all the studied EDAs. For that, we graphically show the evolution of the best individual's fitness value (SEP) in each generation during a typical execution for the six algorithms (average values after 50 executions). In order to analyze the behavior of the algorithms, we plot the best value of SEP in the population at each generation for Cancer and Diabetes (as an example) in Figs. 5-4 and 5-5, respectively. All the algorithms performed 81 generations with the exception of BOA, which only made 50 generations due to its different parameterization.

As can be seen in Figs. 5-4 and 5-5, the evolution of the best value of SEP during the execution is quite similar for all the algorithms. Nevertheless, it is worth to highlight that in Fig. 5-4 BOA converges faster than the others to the lowest (best) values for SEP, but the final value found is similar to that of the other algorithms. Conversely, the slowest convergence is performed by MIMIC$_C$ for the two plotted problems.

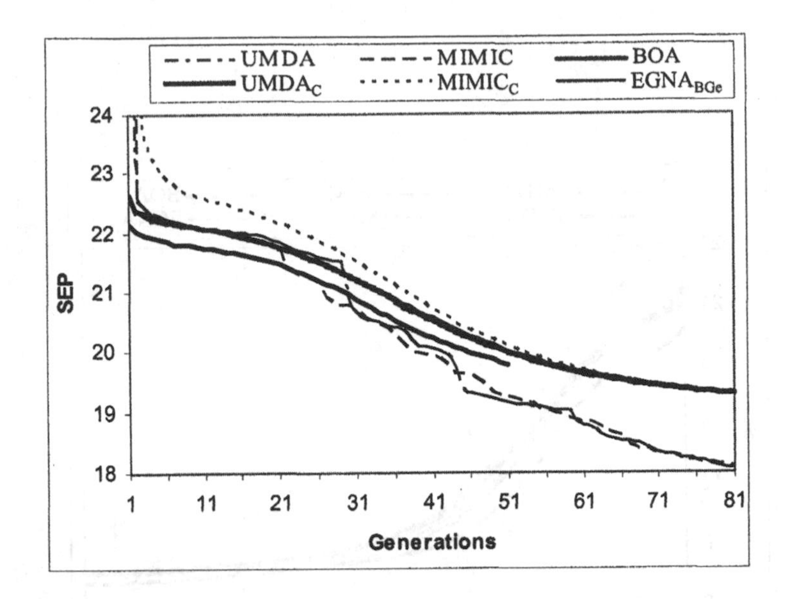

Figure 5-5. Evolution of SEP (best value) for the algorithms in the Diabetes database.

4.3 Comparison Versus Other Works

There exist several interesting works dealing with the problem of training artificial neural networks. Unluckily, many of them use different training sets, and thus their results can not be compared to ours. Nevertheless, some of them study the same problem instances as in this work. In Table 5-7 we compare the best results we have found in our tests versus those of seven different algorithms taken from the literature. Specifically, the compared algorithms are two classical techniques for supervised training such as the back-propagation (BP) and the Levenberg-Marquardt (LM) algorithms, a genetic algorithm (GA), two more GAs hybridized with BP (GABP) and LM (GALM), an evolutionary program (EPNet) and, finally, the so called hybrid training method (HTM_1), which combines genetic operators with BP. The fifth first results are published in (Alba and Chicano, 2004) and in Chapter 6 (Alba and Chicano, 2006), while EPNet and HTM_1 were proposed by Yao and Liu (1997), and Erhard et al. (1998), respectively.

Table 5-7. Our best EDA versus other heuristics in terms of average CEP

	BP	LM	GA	GABP	GALM	EPNet	HTM_1	Best EDA
Cancer	0.91	3.17	16.76	1.43	0.02	1.38	-	1.36 ($MIMIC_C$)
Diabetes	21.76	25.77	36.46	36.46	28.29	22.37	30.11	44.99 (BOA)
Heart	27.41	34.73	41.50	54.30	22.66	16.77	47.08	36.34 ($UMDA_C$)
Thyroid	7.28	2.14	7.28	7.28	7.28	2.12	-	7.28 (BOA)

The best results we have found in this work are highly competitive with some of the best ones in the literature. We have not been able to improve the best-known solution for any problem, but our algorithms outperform some of the compared ones in every problem. The exception is the Diabetes instance, for which the difference between our best solution (obtained by BOA) and GA and GABP is 18.96%. Hence, our EDAs improve the results of other algorithms like LM, GA, GABP, and EPNet for Cancer, GA, GABP, and HTM_l for Diabetes, or GA, GABP, and HTM_l in the case of Heart.

We consider that these are important results, since all our EDAs (which have many different features) have been tested with the same common parameters. It is then expected that our results should be improved in the case of tuning the parameters for each EDA. Anyway, the six studied EDAs improve the results reported for LM and GA by Alba and Chicano (2004) for Cancer, or GA, GABP (both also from Alba and Chicano, 2004), and HTM_l (Erhard et al., 1998) for Heart.

5. CONCLUSIONS

In this chapter, the problem of supervised learning has been faced with EDAs. Specifically, our proposal consists in the study of six different algorithms, belonging to the family of EDAs, for training a multilayer perceptron. We have considered algorithms having univariate, bivariate, or multivariate dependences among variables, and they work over continuous and discrete search spaces. In order to test all these algorithms, we have selected four well-known classification problems, which belong to the area of Medicine: breast cancer diagnosis, diabetes diagnosis in Pima Indians, heart diseases diagnosis, and diagnosis of thyroid hyper- or hypofunction. All our results have been compared versus those of other existing algorithms in the literature.

We conclude in this work that the EDAs working in a continuous space ($UMDA_C$, $MIMIC_C$, and $EGNA_{BGe}$) outperform the EDAs with discretized variables (UMDA, MIMIC, and BOA) both in terms of accuracy (quality of the solutions found) and execution time for this problem. Indeed, the execution time of continuous EDAs can achieve 500 times faster than discrete EDAs in some cases. The reason for the poor performance of discrete EDAs (with respect to the continuous ones) is probably the high number of variables they have to manage for computing the distribution probability, since 16 binary variables (12 for BOA) are used for discretizing the weights of the connections between artificial neurons, as well as the bias terms. In the case of EDAs in continuous domains, one variable (of type *double*) is used for each weight.

Our results were compared to other existing results in the literature, obtained by classical heuristics for this problem, genetic algorithms, memetic algorithms, or evolutionary programming, among others. These comparisons have demonstrated that our EDAs are highly competitive with the compared techniques, being able to improve or almost equalize the results of every compared algorithm in one or more of the studied problems.

The results of this preliminary work are highly promissory, encouraging the authors to deep in this study. Then, as a future work, we are planning to deal with both parallel versions of EDAs and decentralized ones for solving the problem of training neural networks. This way, execution times should be reduced and thus problems with a larger number of variables (weights) could be faced. This will be specially relevant for algorithms using complex models for the probability distribution (multivariate EDAs). Additionally, it is expected that tuning specific parameters for each algorithm will lead us to better results, since we have used the same values for the common parameters of the six EDAs studied in this work in order to make our comparisons as fair as possible.

REFERENCES

Alba, E., Chicano, J. F., 2004, Training neural networks with GA hybrid algorithms, in: *Proc. of Genetic and Evolutionary Computation Conference (GECCO)*, K. Deb, ed., volume 3102 of LNCS, Springer-Verlag, Seattle, Washington, USA, pp. 852-863.

Alba, E., Chicano, J. F., 2006, Genetic algorithms, in: *Metaheuristic Procedures for Training Neural Networks*, E. Alba and R. Martí, eds., Springer, Chapter 6.

Alba, E., Dorronsoro, B., 2005, The exploration/exploitation tradeoff in dynamic cellular evolutionary algorithms, *IEEE Transactions on Evolutionary Computation* **9**(2):126-142.

Bäck, T., Fogel, D., Michalewicz, Z., 1997, *Handbook of Evolutionary Computation*, Oxford University Press, London, U.K.

Baluja, S., 1994, Population-based incremental learning: A method for integrating genetic search based function optimization and competitive learning. Technical Report CMU-CS-94-163, Carnegie Mellon University.

Baluja, S., 1995, An empirical comparison of seven iterative and evolutionary function optimization heuristic. Technical Report CMU-CS-95-193, Carnegie Mellon University.

Bennett, K. P., Mangasarian, O. L., 1992, Robust linear programming discrimination of two linearly inseparable sets, *Optimization Methods and Software*, pp. 23-34.

Bosman, P. A. N., Thierens, D., 1999, Linkage information processing in distribution estimation algorithms, in: *Proc. of the Genetic and Evolutionary Computation Conference (GECCO)*, volume I, Morgan Kaufmann Publishers, Orlando, Florida, USA, pp. 60-67.

Cantú-Paz, E., 2002, Feature subset selection by estimation of distribution algorithms, in: *Proc. of the Genetic and Evolutionary Computation Conference (GECCO)*, Morgan Kaufmann, San Francisco, CA, pp. 303-310.

Cantú-Paz, E., 2003, Pruning neural networks with distribution estimation algorithms, in: *Proc. of Genetic and Evolutionary Computation Conference (GECCO)*, E. Cantú-Paz et al., eds., volume 2723 of LNCS, Springer-Verlag, Illinois, Chicago, USA, pp. 790-800.

Cotta, C., Alba, E., Sagarna, R., Larrañaga, P., 2002, Adjusting weights in artificial neural networks using evolutionary algorithms, in: *Estimation of Distribution Algorithms. A New Tool for Evolutionary Computation*, P. Larrañaga and J. A. Lozano, eds., Kluwer.

Dawn, T., 1994, Neural computing makes its mark in science, *Scientific Computing* **3**:25-30.

De Bonet, J. S., Isbell, C. L., Viola P., 1997, MIMIC: Finding optima by estimating probability densities, in: *Advances in Neural Information Processing Systems*, M. Mozer et al., eds., volume 9, The MIT Press, page 424.

Detrano, R., Janosi, A., Steinbrunn, W., Pfisterer, M., Schmid, J., Sandhu, S., Guppy, K., Lee, S., Froelicher, V., 1989, International application of a new probability algorithm for the diagnosis of coronary artery disease, *American Journal of Cardiology*, pp. 304-310.

Erhard, W., Fink, T., Gutzmann, M. M., Rahn, C., Doering, A., Galicki, M., 1998, The improvement and comparison of different algorithms for optimizing neural networks on the MasPar MP-2, in: *Neural Computation*, M. Heiss, ed., ICSC Press, pp. 617-623.

Fujita, H., Katafuzhi, T., Uehara, T., Nishimura, T., 1992, Neural network approach for the computer-aided diagnosis of coronary artery diseases in nuclear medicine, in: *International Joint Conference on Neural Networks*, Baltimore, USA, pp: 215-220.

Galic, E., Höhfeld, M., 1996, Improving the generalization performance of multi-layer-perceptrons with population-base incremental learning, in: *Parallel Problem Solving from Nature IV (PPSN)*, vol. 1141 of LNCS, Springer-Verlag, pp. 740-750.

Geiger, D., Heckerman, D., 1994, Learning Gaussian networks, Technical Report MST-TR-94-10, Microsoft Advanced Technology Division, Microsoft Corp., Seattle, Washington.

Gennari, J. H., Langley, P., Fisher, D., 1989, Models of incremental concept formation, *Artificial Intelligence* **40**:11-61.

Larrañaga, P., Etxeberria, R., Lozano, J. A., Peña, J. M., 1999, Optimization by learning and simulation of Bayesian and Gaussian networks, Technical Report KZZA-IK-4-99, Dept. of Computer Science and Artificial Intelligence, University of the Basque Country.

Larrañaga, P., Etxeberria, R., Lozano, J. A., Peña, J. M., 2000, Optimization in continuous domain by learning and simulation of Gaussian networks, in: *Proc. of the Genetic and Evolutionary Computation Workshop Program*, A. S. Wu, ed., Morgan Kaufmann, Las Vegas, Nevada, USA, pp. 201-204.

Larrañaga, P., Lozano. J. A., 2002, *Estimation of Distribution Algorithms. A New Tool for Evolutionary Computation*, Kluwer Academic Publishers.

Mangasarian, O. L., Setiono, R., Wolberg, W. H., 1990, Pattern recognition via linear programming: Theory and application to medical diagnosis, in: *Large-Scale Numerical Optimization*, T. F. Coleman, and Y. Li, eds., SIAM Publications, Philadelphia, pp. 22-31.

Maxwell, B., Anderson, S., 1999, Training hidden Markov models using population-based learning, in: *Proc. of the Genetic and Evolutionary Computation Conference (GECCO)*, W. Basnzhaf et al., eds., Morgan Kaufmann, Orlando, Florida, USA, pp. 944.

McClelland, J. L., Rumelhart, D. E., 1996, *Parallel Distributed Processing: Explorations in the Microstructure of Cognition*. The MIT Press.

Mühlenbein, H., 1997, The equation for response to selection and its use for prediction, *Evolutionary Computation* **5**(3):303-346.

Mühlenbein, H., and Höns, R., 2005, The estimation of distributions and the minimum relative entropy principle, *Evolutionary Computation* **13**(1):1-27.

Mühlenbein, H., Mahnig, T., 2000, Evolutionary optimization using graphical models, *New Generation Computing* **18**(2):157-166.

Mühlenbein, H., Mahnig, T., Ochoa, A., 1999, Schemata, distributions and graphical models in evolutionary optimization, *Journal of Heuristics* **5**:215-247.

Mühlenbein, H., Paaß, G., 1996, From recombination of genes to the estimation of distributions I. Binary parameters, in: *Parallel Problem Solving from Nature IV (PPSN)*, H.-M Voigt, et al., eds., LNCS 1141, Springer-Verlag, pp. 178-187.

Mühlenbein, H., Voigt, H. M., 1996, Gene poll recombination in genetic algorithms, in: *Metaheuristics: Theory and Applications*, pp. 53-62.

Pelikan, M., Goldberg, D. E., 2000, Research on the Bayesian optimization algorithm, in: *Proc. of the Genetic and Evolutionary Computation Workshop Program*, A. S. Wu, ed., Morgan Kaufmann, Las Vegas, Nevada, USA, pp. 212-215.

Pelikan, M., Goldberg, D. E., Cantú-Paz, E., 1999a, BOA: The Bayesian optimization algorithm, in: *Proc. of the Genetic and Evolutionary Computation Conference (GECCO)*, W. Banzhaf et al., eds., volume 1, Morgan Kaufmann, Orlando, Florida, pp. 525–532.

Pelikan, M., Goldberg, D. E., Cantú-Paz, E., 2000a, Bayesian optimization algorithm, population sizing, and time convergence, in: *Proc. of the Genetic and Evolutionary Computation Conference (GECCO)*, D. Whitley et al., eds., Morgan Kaufmann, Las Vegas, Nevada, USA, pp. 275-282.

Pelikan, M., Goldberg, D. E., Cantú-Paz, E., 2000b, Linkage problem, distribution estimation and Bayesian networks, *Evolutionary Computation* **8**(3):311-340.

Pelikan, M., Goldberg, D. E., Lobo, F. G., 1999b, A survey of optimization by building and using probabilistic models. Technical Report No. 99018, Urbana, IL: University of Illinois Genetic Algorithms Laboratory (IlliGAL).

Pelikan, M., Mühlenbein, H., 1999, The bivariate marginal distribution algorithm, in: *Advances in Soft Computing-Engineering Design and Manufacturing*, pp. 521-535.

Prechelt, L., 1994, Proben1—A set of neural network benchmark problems and benchmarking rules. Technical Report 21/94, University of Karlsruhe.

Rosenblatt, F., 1962, *Principles of Neurodynamics: Perceptrons and the Theory of Brain Mechanisms*. Spartan Books, New York.

Rumelhart, D., Hinton, G., Williams, R., 1986, Learning representations by backpropagation errors, *Nature* **323**:533-536.

Simpson, R., 1992, Biological pattern recognition by neural networks, *Marine Ecology Progress Series* **79**:303-308.

Smith, J. W., Everhart, J. E., Dickson, W. C., Knowler, W. C., Johannes, R. S., 1988, Using the ADAP learning algorithm to forecast the onset of diabetes mellitus, in: *Proc. of the Twelfth Symposium on Computer Applications in Medical Care*, IEEE Computer Society Press, Washington D.C., pp. 261-265.

Soto, M., Ochoa, A., Acid, S., de Campos, L. M., 1999, Introducing the polytree approximation of distribution algorithm, in: *Second Symposium on Artificial Intelligence. Adaptative System CIMAF*, Editorial Academia, La Habana, Cuba, pp. 360-367.

Wolberg, W. H., 1990, Cancer diagnosis via linear programming, *SIAM News* **23**(5):1-18.

Wolberg, W. H., Mangasarian, O. L., 1990, Multisurface method of pattern separation for medical diagnosis applied to breast cytology, *Proc. of the National Academy of Sciences* **87**:9193-9196.

Yao, X., Liu, Y., 1997, A new evolutionary system for evolving artificial neural networks, *IEEE Transactions on Neural Networks* **8**(3):694-713.

Zhang, B. T., Cho, D. Y., 2000, Evolving neural tress for time series prediction using Bayesian evolutionary algorithms, in: *Proc. of the First IEEE Workshop on Combinations of Evolutionary Computation and Neural Networks (ECNN)*, X. Yao, ed., San Antonio, Texas, USA, pp. 17-23.

Chapter 6

GENETIC ALGORITHMS

Enrique Alba and Francisco Chicano
Department of Languages and Computer Science, University of Málaga, Spain

Abstract: In this chapter we describe the basics of Genetic Algorithms and how they can be used to train Artificial Neural Networks. Supervised training of Multilayer Perceptrons for classification problems is considered. We also explain how the Genetic Algorithm can be hybridized with other algorithms and present two hybrids between it and two classical algorithms for the neural network training: Backpropagation and Levenberg-Marquardt. Several experiments over a set of six applications in the context of Bioinformatics are performed comparing the Genetic Algorithm, its hybrids, and the classical algorithms mentioned above. The testbed has been chosen from PROBEN1: breast cancer, diabetes, heart disease, gene, soybean, and thyroid. The results show that the genetic algorithm hybridized with Levenberg-Marquardt is a serious competitor for standard approaches.

Key words: Neural networks; genetic algorithms; hybridization; bioinformatics; medical applications; Proben1.

1. INTRODUCTION

The interest in making research on Artificial Neural Networks (ANNs) resides in the appealing properties that ANNs exhibit: adaptability, learning capability, and ability to generalize. Nowadays, ANNs are receiving a great deal of attention from the international research community as it is shown in the large number of studies concerning training, structure design, and real world applications, ranging from classification to robot control or vision (Alander, 1994; Lisboa et al., 2000).

The neural network training task is a capital process in supervised learning, in which a pattern set made up with pairs of inputs plus expected outputs is known beforehand. This set of patterns is used to compute (find out) the set of weights that makes the ANN to learn it. To achieve this goal, the algorithm must modify the weights of the neural network in order to get the desired output for a given input, usually in an iterative manner, until a minimum error between the current and the expected output is attained (i.e., the pattern set is learned and the ANN is trained).

One of the most popular training algorithms in the domain of neural networks is the Backpropagation technique (generalized delta rule) (Rumelhart et al., 1986), a gradient-descent method. Other techniques, usually based on metaheuristic optimization engines such as Evolutionary Algorithms (EAs), have been also applied to the training problem in recent years (Cotta et al., 2001; Cantú-Paz, 2003), in the quest for efficient algorithms able of avoiding the local minima in such a complex problem. Although training is a main issue in ANN design, many other works exist addressing the evolution of the layered structure of the ANN or even the elementary behavior of the neurons composing the ANN. For example, in Alba et al. (1993) a definition of neurons, layers, and the associated training problem is faced by using parallel genetic algorithms; also, in Yao and Liu (1997) the architecture of the network and the weight values are evolved by using the EPNet evolutionary system. The literature on this topic is really vast, the work of Yao (1999) represents an excellent starting point to get acquired in the research of ANN training.

In this chapter our main objective is to describe and illustrate how Genetic Algorithms (GAs) can be used to train ANNs. We address two fundamental tasks in the resolution of any problem with GAs: the representation of the solutions and the design of the fitness function. We also show how GAs can be hybridized with other local search methods in order to outperform the results. At the same time, we want to perform a standard presentation of results that promotes and makes easy future comparisons. This sounds common sense, but it is not frequent that authors follow commonly accepted procedures for comparisons such as the structured Prechelt's set of recommendations (Prechelt, 1994), a "de facto" standard for many ANN researchers.

For the empirical study we have selected a benchmark from the field of Bioinformatics, composed of six classification problems: diagnosis of breast cancer, diagnosis of diabetes in Pima Indians, diagnosis of heart disease, splice junctions detection in nucleotide sequences, recognition of different diseases of soybeans, and diagnosis of thyroid hyper- or hypofunction. All of them represent well-known benchmark problems publicly available at:

- `http://www.ics.uci.edu/~mlearn/MLRepository.html`
- `ftp://ftp.ira.uka.de/pub/neuron/proben1.tar.gz`

In our empirical study we compare the GA and its associated hybrid algorithms with two classical algorithms in the ANN domain such as Backpropagation (BP) algorithm and Levenberg-Marquardt (LM) approach.

The remainder of the chapter is organized as follows. Section 2 gives a brief description of the foundations of neural networks. Next, we present the genetic algorithms in Section 3. The following section discusses how GAs can be applied to the ANN supervised training, specifying the mechanisms used for representing solutions and evaluating their quality, and some concepts of hybridization. After discussing the algorithm, the representation, and the associated fitness function, we present in Section 5 the experimental study. Finally, we summarize the conclusions and future work in Section 6.

2. ARTIFICIAL NEURAL NETWORKS

Artificial Neural Networks are computational models naturally performing a parallel processing of information (McClelland and Rumelhart, 1986). Essentially, an ANN can be defined as a pool of simple processing units (neurons) which communicate among themselves by means of sending analog signals. These signals travel through weighted connections between neurons. Each of these neurons accumulates the inputs it receives, producing an output according to an internal activating function. This output can serve as an input for other neurons, or it can be a part of the network output. In Figure 6-1 we can see a neuron in detail.

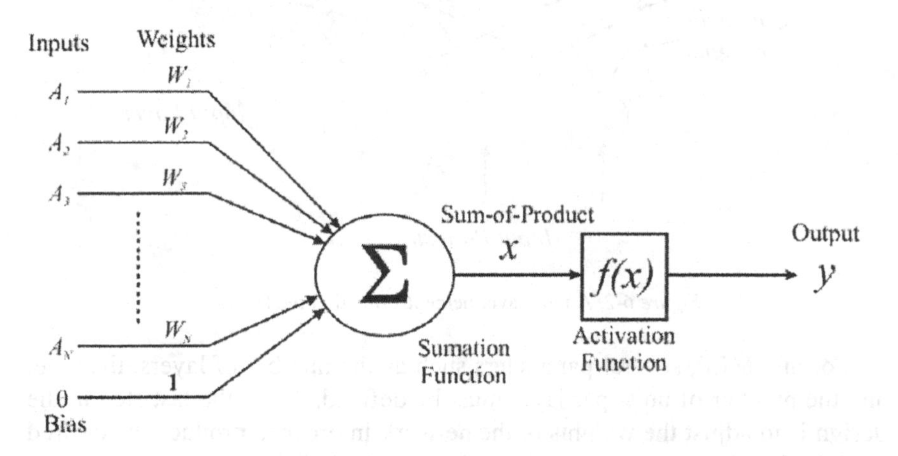

Figure 6-1. An artificial neuron.

There is a set of important issues involved in the ANN design process. As a first step, the architecture of the network has to be decided. Initially, two major options are usually considered: feedforward networks and recurrent networks (additional considerations regarding the order of the ANN exist, but they are out of our scope). The feedforward model comprises networks in which the connections are strictly feedforward, i.e., it is possible to number the neurons in such a way that for each connection (i, j) it holds $i<j$. The recurrent model defines networks in which feedback connections are allowed, thus inducing complex dynamical properties in the ANN. In this article we will concentrate on the first model: the feedforward networks. To be precise, we will consider the so-called *multilayer perceptron* (MLP) (Rosenblat, 1962), in which units are structured into ordered layers, and connections are allowed only between adjacent layers in an input-to-output sense (see Figure 6-2).

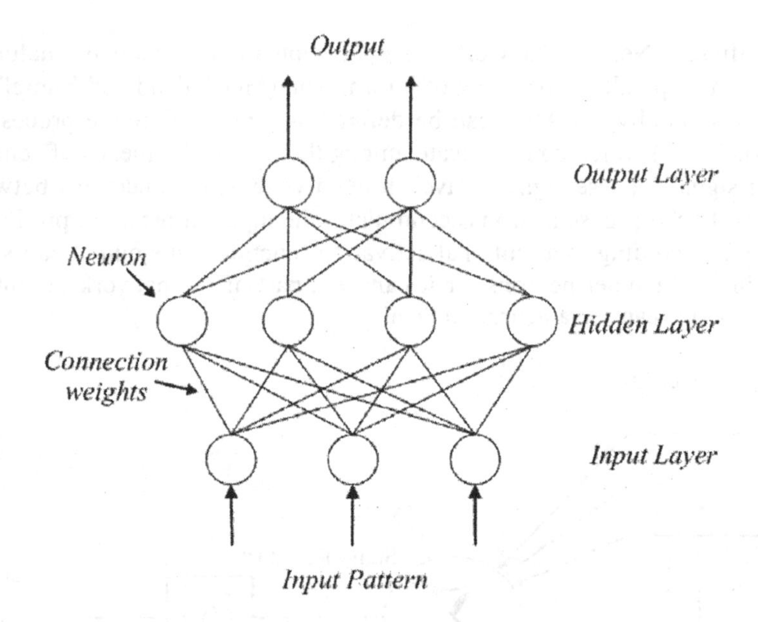

Figure 6-2. A multilayer perceptron with three layers.

For any MLP, several parameters such as the number of layers, the type, and the number of units per layer must be defined. Then, the last step in the design is to adjust the weights of the network in order to produce the desired output when its corresponding input is presented. This process is known as training the ANN. Network weights comprise both the previously mentioned connection weights as well as bias terms for each unit. The latter can be viewed as the weight of a constant saturated input that the corresponding unit always receives. As initially stated, we will focus on the learning situation known as supervised training, in which a set of target input/output patterns is available. The input and output of the network are both real valued vectors in our case.

3. GENETIC ALGORITHMS

Genetic Algorithms are a kind of Evolutionary Algorithm (Bäck, 1996). They are loosely based on the principles of natural evolution, namely adaptation and survival of the fittest. These techniques have been shown to be very effective in solving hard optimization problems like training ANNs, i.e., problems in which gradient-descent techniques get trapped into local minima, or are fooled by the complexity and/or non-differentiability of the search space.

The underlying idea in Genetic Algorithms (GAs) is that of using multiple *individuals*, each one encoding a tentative solution to the problem considered. The set of all individuals is the so-called *population.* Each individual has an associated quality value (the fitness value). In every algorithm step, known as *generation*, the individuals in the current population are manipulated by several variation operators to form a new population (the next generation). In GAs the individuals usually represent the tentative solutions by means of binary strings. However, other representations can be used (Eshelman, 2000).

```
t := 0;
P(0) := Generate ();
Evaluate (P(0));
while not StopCriterion do
    P'   (t) := Select (P(t));
    P''  (t) := Recombination (P'(t));
    P''' (t) := Mutation (P''(t));
    Evaluate (P'''(t));
    P(t+1)   := Replace (P(t), P'''(t));
    t := t+1;
endwhile;
```

Figure 6-3. Pseudocode of a Genetic Algorithm.

We present in Figure 6-3 the detailed operations in each step of a GA. First, an initial population has to be generated randomly or by using a seeding algorithm. This initial population is evaluated and then the algorithm enters the main loop. Next, several individuals are selected from the current population according to their fitness values. Many selection procedures are currently in use, one of the simplest being Holland's original fitness-proportionate selection (or *Roulette selection*), where individuals are selected with a probability proportional to their relative fitness. This ensures that the expected number of times an individual is chosen is approximately proportional to its relative performance in the population. Thus, high-fitness ("good") individuals stand a better chance of "reproducing", while low-fitness ones are more likely to disappear (see the schema theorem in Holland, 1975).

The GA uses the selected individuals to create new ones by applying variation operators such as *recombination* and *mutation* operators. The recombination operator combines two or more individuals (solutions) to form new ones. The idea behind the recombination is that the offsprings can get the "good" characteristics of its parents. There are several recombination operators that can be used on binary representations. The Single-Point Crossover (SPX) takes two individuals and randomly selects a position in the two strings. Then, it swaps the substrings at the right of this position to form two new individuals. The Double-Point Crossover (DPX) selects two positions in the string and swaps only the substring between these positions in the two parents. Another well-known recombination operator in the field of GAs is the Uniform Crossover (UX). In this operator each bit of the string is randomly taken from one parent. Many more types of recombination exist for integer, permutation based, and real representations. For a detailed study of them see Booker et al. (2000).

The mutation operator perturbs one individual with a low probability, allowing new characteristics to appear in the population. The mutation operator used in canonical GAs is Bit-Flip mutation. This operator flips each bit in the string with a given probability p_m, usually $p_m=1/length$. By this way, one bit is changed in each individual on average (Bäck et al., 2000).

After recombination and mutation the new individuals are evaluted using the fitness function and the next population is created by means of the *replacement operator*. This last operator uses the new generated individuals and the current population to build up the new one. According to the replacement operator we distinguish two kinds of GAs: *generational* and *steady-state*. In generational GAs all the population is replaced with new individuals. In steady-state GAs only one (or a few) new individual is created, and it usually replaces the worst one in the population (always or only if it is better). The previous operations are repeated until a stopping criterion is fulfilled, such as reaching a maximum number of steps or finding an optimal solution.

4. GENETIC ALGORITHMS FOR ANN TRAINING

We can find some works in the last years in which GAs are used to optimize some aspects of the neural networks. Some works have a theoretical approach, and optimize new neural network models. For example, in Moon and Kong (2001) a new ANN, called Block-Based Neural Network (BBNN) is presented. A GA is used for optimizing the structure and connection weights. In Leung et al. (2003a) an ANN with switches is proposed. The network structure is controlled by the switches and the input-output relationship is controlled by the weights. An improved GA optimizes both the structure and the weights. In Tarca et al. (2004) monotonic-concave neural networks are trained by means of an evolutionary algorithm which combines genetic and hill-climbing search (GA-GHC). Gilson et al. (2003) use a GA and a Tabu Search to train recurrent pulsed networks.

On the other hand, we find some works focused on the application of GAs to train ANNs. Pattnaik et al. (2005) use a GA to optimize several parameters of a neural network. The neural network is then used to calculate the resonant frequency of a coaxially-fed tunable rectangular microstrip-patch antenna. In Ling et al. (2003) a GA is used to train a newly proposed model of neural network. The network is evaluated with the XOR problem and the sunspot number forecasting problem. After that, two applications are tackled: short-term daily home electric load forecasting and pattern recognition. In the work of Ku et al. (1999), GAs are used to train a recurrent neural network solving the long-term dependency problem. A cellular GA with a simple learning method embedded using the Lamarckian mechanism is proposed as a promissory technique to improve the prediction and classification capability of recurrent ANNs. Other applications of neural networks trained with GAs are: detection of defects in conducting pieces (Xiaoyun et al., 2004), recognition of handwritten digits (Leung et al., 2004), sinter's burning through point process (Cheng, 2004), modelling of an electro-discharge machining technique (Wang et al., 2003), hexapod leg control (Parker and Lee, 2003), classification of sperm cells according to their chromosomic content (Kuri-Morales et al., 2003), speech recognition of cantonese commands (Leung et al., 2003b), power quality disturbance classification and recognition (Gengyin et al., 2003), and direct torque control (Cao et al., 2003).

The main reason for the large amount of works training neural networks with GAs is their ability to escape from local optima. The search features of GAs contrast with those of classical algorithms like Backpropagation in that they are not trajectory-driven, but population-driven (Blum and Roli, 2003). In trajectory methods the points in the search space are visited individually, describing a trajectory in the search space. Population-based methods modify in each step a set of solutions, performing an evolution of a set of points in the search space. In this way, a GA is expected to avoid local optima frequently by promoting exploration of the search space, in opposition to the exploitative trend usually allocated to local search algorithms.

In this section we tackle three different issues that must be taken into account when a GA is tuning to work with any problem, in particular, for ANN training: the representation of the solutions, the design of a fitness function, and the hybridization. The first two are required steps when solving a problem with GAs. However, the hybridization is not mandatory but it can outperform the results, as we will see in the experimental section.

4.1 Representation

One of the first topics that must be addressed to solve a problem with any algorithm is the representation of the solutions. This is a very important step in tailoring a general search sheet as a GA for the problem at hand. In our case, a solution is composed of the values of all the weights and biases (variables) in the ANN. Let k be the total number of variables in the ANN. These values can be included in a real vector where each weight (and bias) is allocated always in the same position. This k-dimensional vector could be codified in different ways. In the present analysis we will consider a linear encoding of the vector, i.e., firstly, each variable is conveniently encoded in an algorithm-dependent way; subsequently, the codified vector is constructed by concatenating the encoding of each variable into a linear string. This linear encoding of the variables raises a question: the best distribution of the variables within the string. This distribution is important in connection with the particular recombination operator used in the GA. In fact, if this operator breaks the strings into large blocks using them as units for exchange (like SPX), this distribution might be relevant. On the contrary, using a recombination operator that breaks the strings into very small chunks (like UX) makes the distribution less relevant. A good piece of advice is grouping together the input weights and bias for each unit (Cotta et al., 2001; Alba, 1993). In this way, the probability of transmitting them as a block is increased.

In the GA the encoding of solutions is approached via binary strings. More precisely, m bits are used to represent each single variable; subsequently, the k m-bit segments are concatenated into an l-bit binary string, where $l = k \cdot m$ (Figure 6-4). This encoding of the network variables raises a number of issues. Two of them are the choice of m and the encoding mechanism for the individual variables, i.e., pure binary, Gray-coded numbers, magnitude-sign, etc. In this work we use 16-bit pure binary encoding. The integer value corresponding to an m-bit segment is mapped linearly into a pre-established real interval and the result is the weight value.

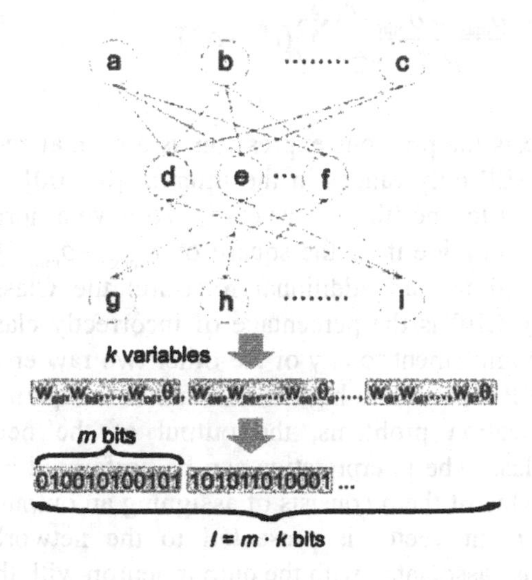

Figure 6-4. Binary representation of the network variables.

4.2 Fitness Function

Now, we discuss the alternatives for the fitness function in order to evaluate the quality of the solutions. In the ANN domain the Mean Squared Error (MSE) is a popular measure that evaluates the ANN output error with respect to the expected output. This term measures the proximity of the current output to the desired output. We can compute this error term just for one single pattern or for a set of patterns. In the latter case, the MSE is the average value of the separate MSE values. The expression for this global MSE is

$$MSE = \frac{\sum_{p=1}^{P} \sum_{i=1}^{S} (t_i^p - o_i^p)^2}{P \cdot S} \quad (1)$$

where t_i^p and o_i^p are, respectively, the i-th components of the expected vector and the current output vector for the pattern p, S is the number of output neurons, and P is the number of patterns. This MSE value is closely related to the Squared Error Percentage (SEP) proposed by Prechelt (1994). If the output values of the network output neurons are in the range $[o_{min}, o_{max}]$, the expression for the SEP is

$$SEP = 100 \cdot \frac{o_{\max} - o_{\min}}{P \cdot S} \sum_{p=1}^{P} \sum_{i=1}^{S} (t_i^p - o_i^p)^2 \tag{2}$$

Prechelt suggests the previous expression as a normalized version of the MSE. However, SEP only ranges in the interval [0, 100] if the difference $o_{\max} - o_{\min}$ is equal to one (this is our case). To have a normalized version of MSE we need to divide it by the square of $o_{\max} - o_{\min}$. In classification problems, we could use an additional measure: the Classification Error Percentage (CEP). CEP is the percentage of incorrectly classified patterns, and it is a usual complement to any of the other two raw error values (MSE or SEP), since CEP reports in a high-level manner the quality of the trained ANN. In classification problems, the output of the network must be interpreted as a class. The interpretation can be performed in different ways (Prechelt, 1994). One of them consists of assigning an output neuron to each class. When an input vector is presented to the network, the network response is the class associated with the output neuron with the largest value. We utilize this method here, that is known as *winner-takes-all.* The objective of the network is to classify all the patterns correctly, that is, to obtain 0% of CEP. We could use CEP as a function to be minimized. However, if two networks happen to have the same CEP for the pattern set, we could distinguish between them more precisely by looking at their output values, since the output of one of them can be nearer to the desired output than the other. For this reason, the SEP value can possibly be a better guide for the search. In our study, the fitness function to be minimized is the SEP value for the pattern set. We tackle a maximization approach that considers the inverse of the SEP as fitness, that is,

$$f(\mathbf{x}) = \frac{1}{SEP(\mathbf{x})} \tag{3}$$

The previous fitness function is undefined when the ANN reaches a SEP of zero, that is, when the current output is equal to the desired output. In that case an alternative fitness function (not used in this work) can be

$$f(\mathbf{x}) = \frac{1}{1 + SEP(\mathbf{x})} \tag{4}$$

4.3 Hybridization

Hybridization is almost "a must" in complex applications if one expects to get algorithms that are efficient and accurate at the same time. In our study, hybridization refers to the inclusion of problem-dependent knowledge in a general search template (Cotta and Troya, 1998; Davis, 1991). We can distinguish between two kinds of hybridization: strong and weak hybridization. In the former, the knowledge is included using specific operators or representation. In the latter, several algorithms are combined somehow. In this last case, an algorithm can be used to improve the results of another one separately or it can be used as an operator of the other (Talbi, 2002).

The hybrid algorithms that we use in this paper are combinations of two algorithms (weak hybridization), where one of them acts as an operator inside the other. We combine a GA with the classical Backpropagation algorithm (GABP), and a GA with Levenberg-Marquardt (GALM). In both cases the problem-specific algorithm (BP and LM) replaces the mutation operator of the primary search template (GA). Therefore, we can say that GAxx is a GA in which the mutation has been replaced by the "xx" algorithm that is applied with probability p_t (Figure 6-5). We describe BP and LM briefly in the Appendices A and B, respectively.

```
t := 0;
P(0)  := Generate ();
Evaluate (P(0));
while not StopCriterion do
     P'   (t)  := Select (P(t));
     P''  (t)  := Recombine (P'(t));
     P'''(t)  := Hybrid_Op_XX (P''(t));
     Evaluate (P'''(t));
     P(t+1)    := Replace (P(t), P'''(t));
     t  := t+1;
endwhile;
```

Figure 6-5. Pseudocode of a hybrid algorithm GAxx.

With respect to the representation in these hybrid algorithms, different encoding schemes are used in the different phases of the training. BP and LM use real vectors representing the weights of the ANN while the GA uses a binary string in the representation. Since an exchange of information (tentative solutions) is needed from the GA to BP (or LM), a translation between the binary string and the real valued vector is performed. In the opposite sense, when the network is trained with BP or LM it is necessary a translation from real valued vector to binary string (usually with a loss of precision).

5. EXPERIMENTAL ANALYSIS

In this experimental section we will compare the GA and the two hybrid algorithms (GABP and GALM) against the two classical algorithms used in the hybridization. In this way, we can study the advantages of the evolutionary approach and the emergent properties resulting after the combination of GA and the classical algorithms.

We address six classification problems from the domain of Bioinformatics. The instances solved here belong to the PROBEN1 set of problems: Cancer, Diabetes, Heart, Gene, Soybean, and Thyroid. These instances can also be found at the UCI Machine Learning Repository, but in PROBEN1 they have been preprocessed with a set of rules proposed by Prechelt (1994). The goal of this preprocessing is to get a standard way of comparing algorithms against the results of other authors that also follow the same rules. We now briefly detail the instances:

- **Cancer (BC)**: Diagnose breast cancer. The problem consists of classifying a tumor as either benign or malignant based on cell descriptions gathered by microscopic examination. There are 699 test cases that were obtained by Dr. William H. Wolberg at the University of Wisconsin Hospitals, Madison (Bennett and Mangasarian, 1992; Mangasarian et al., 1990; Wolberg, 1990; Wolberg and Mangasarian, 1990).
- **Diabetes (DI)**: Diagnose diabetes of Pima Indians. Based on personal data and the results of medical examinations, the task is to decide whether a Pima Indian individual is diabetes positive or not. There are 768 test cases from the National Institute of Diabetes and Digestive and Kidney Diseases by Vincent Sigillito (Smith et al., 1988).
- **Heart (HE)**: Predict heart disease. The problem consists of deciding whether at least one of four major vessels is reduced in diameter by more than 50%. This decision is made based on personal data and results of medical examinations. There are 920 test cases from four different sources: Hungarian Institute of Cardiology in Budapest (Andras Janosi, M.D.), University Hospital of Zurich in Switzerland (William Steinbrunn, M.D.), University Hospital of Basel in Switzerland (Mathhias Pfisterer, M.D.), V.A. Medical Center of Long Beach and

Cleveland Clinic Foundation (Robert Detrano, M.D., Ph.D.) (Detrano et al., 1989; Gennari et al., 1989).

- **Gene (GE)**: Detect intron-exon boundaries in nucleotide sequences. From a window of 60 DNA sequence elements the goal is to decide whether the middle is either an intron-exon boundary, an exon-intron boundary, or none of them. There are 3175 test cases donated by G. Towell, M. Noordewier, and J. Shavlik and obtained from *GenBank 64.1* (ftp://genbank.bio.net) (Noordewier et al., 1991).
- **Soybean (SO)**: Recognize 19 different diseases of soybeans. The discrimination is done based on a description of the bean and the plant plus information about the history of the plant life. There are 683 test cases from Ming Tan and Jeff Schlimmer (Tan and Eshelman, 1988).
- **Thyroid (TH)**: Diagnose thyroid hyper- or hypofunction. Based on the patient query data and patient examination data, the task is to decide whether the patient thyroid has overfunction, normal function, or underfunction. There are 7200 test cases donated by Randolf Werner and obtained from Daimler-Benz (Schiffmann et al., 1992).

The structure of the MLP used for all the problems accounts for three layers (input-hidden-output) having six neurons in the hidden layer. The number of neurons in the input and output layers depends on the concrete problem instance. The activating function of the neurons (including those of the output layer) is the sigmoid function. Table 6-1 summarizes the network architecture for each instance.

Table 6-1. MLP architecture and patterns distribution for all the instances

	Architecture			**#Patterns**		**Weights**
Instances	**in**	**hidden**	**out**	**Train**	**Test**	**and biases**
Cancer	9	6	2	525	174	74
Diabetes	8	6	2	576	192	68
Heart	35	6	2	690	230	230
Gene	120	6	3	2382	793	747
Soybean	82	6	19	513	170	631
Thyroid	21	6	3	5400	1800	153

To evaluate an ANN, we split the pattern set into two subsets: the training one and the test one. The ANN is trained by using the training pattern set, and then it is evaluated on the unseen test pattern set. The training set for each instance is approximately made of 75% of the examples, while the other 25% constitutes the test set. The exact number of patterns for each instance is presented in Table 6-1 to ease future comparisons.

After presenting the problems, we now turn to describe the parameters of the algorithms, as shown in Table 6-2. The hybrid algorithms use the same parameters as their elementary components. However, the mutation operator of the GA is not applied but it is replaced by BP or LM, respectively. BP and LM are applied with an associated probability p_t, just like any other classical genetic operator. When applied, BP (LM) only performs one single epoch.

The BP and LM algorithms use real representation for the variables of the network, while the GA and the hybrid algorithms use binary vectors. These vectors are 16-bit length and represent a real value in the interval [-1,+1]. The weights (variables) associated to input links for a neuron are placed contiguously in the chromosome. The GA uses a steady-state strategy (one new solution in every step is computed). Fifty independent runs are performed to assess the statistical significance of the results. The machines used are Pentium 4 at 2.8GHz with 512MB of RAM running Linux (kernel version 2.4.19-4GB). All these algorithms were written in Java (JRE version 1.4.2_03).

In the following subsections we present and analyze the results obtained after the application of the five algorithms to the six instances. The next section analyzes CEP and SEP obtained for each pair instance-algorithm. Then, we study the computational effort of the algorithms in Section 5.2 and, finally, we compare our results with the literature in Section 5.3.

Table 6-2. Parameters of the algorithms

Algorithms	Parameters	BC	DI	HE	GE	SO	TH
BP	Epochs	1000	1000	500	19	18	63
	η	10^{-2}	10^{-2}	10^{-3}	10^{-4}	10^{-3}	10^{-4}
	α	0.0	0.0	0.0	0.0	0.0	0.0
LM	Epochs	1000	1000	500	19	18	63
	μ	10^{-3}	10^{-3}	10^{-3}	10^{-3}	10^{-3}	10^{-3}
	β	10	10	10	10	10	10
GA	Population size	64					
	Selection	Roulette (2 inds.)					
	Recombination	Single point (p_c=1.0)					
	Mutation	Bit-Flip (p_m=1/*length*)					
	Replacement	Elitist					
	Stop criterion	1000 generations (1064 evals.)					
GAxx	Epochs of "xx"	1	1	1	1	1	1
	p_t	1.0	1.0	0.5	0.019	0.018	0.063

5.1 Training Error and Classification

In this section we report the mean and the standard deviation of CEP and SEP for the test pattern set. In Table 6-3 and Figure 6-6 we show the results of CEP; Table 6-4 and Figure 6-7 show the results of SEP. Best results are marked in boldface.

For the Cancer instance the best results of CEP are obtained (from best to worst) by GALM, BP, and GABP. They are followed by LM and, finally, by the GA. This last is not a surprising fact, since the GA performs a rather explorative search in this kind of problems. Observing SEP, the lowest (best) values are reached by BP, while the GA gets the highest one as expected. SEP and CEP values are related to each other, that is, a lower value of SEP usually implies a lower value for CEP, but in some cases this relationship is not direct. For example, GABP and GALM have a larger value of SEP than LM, but they have a lower value of CEP.

In Diabetes, BP shows the highest accuracy of CEP, followed by LM and GALM. In this case, GA and GABP obtain the same CEP. In the two algorithms all the trained ANNs classify all the patterns into the same class (which is wrong). The best SEP is here obtained by BP. On the other hand, LM, GA, and GABP get the worst values of SEP. Comparing Figures 6-6 and 6-7 we see a similar contour of the bars except for the LM algorithm.

Table 6-3. Results of CEP in ANN training

CEP(%)		BP	LM	GA	GABP	GALM
BC	Avg.	0.84	3.53	16.64	1.90	**0.01**
	Dev.	0.29	1.23	5.89	5.65	0.08
DI	Avg.	**21.75**	24.75	36.46	36.46	28.21
	Dev.	0.32	2.79	0.00	0.00	0.98
HE	Avg.	27.67	35.13	46.68	54.22	**22.70**
	Dev.	1.43	3.56	16.08	20.25	0.88
GE	Avg.	49.76	15.51	58.75	13.89	**13.51**
	Dev.	45.97	3.11	34.05	19.23	4.78
SO	Avg.	97.74	**28.11**	94.51	96.12	59.40
	Dev.	7.71	12.87	8.67	8.13	23.28
TH	Avg.	7.28	**2.14**	7.28	7.28	7.28
	Dev.	0.00	1.13	0.00	0.00	0.00

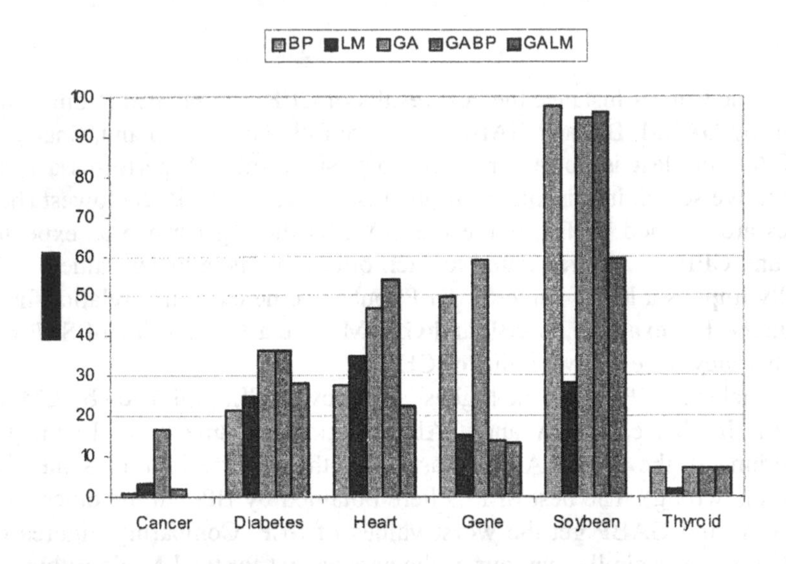

Figure 6-6. Comparison of CEP among the algorithms.

For the Heart problem, GALM gets the best CEP, followed by BP; and GABP gets the worst one. We found here approximately the same behavior as for the two previous instances. In this problem we can see in Figure 6-6 a clear advantage in accuracy of a hybrid (GALM) over its two separate components (GA, LM): a new behavior emerges from their combination. In SEP, GALM gets the best value, followed by BP; and LM reaches the worst SEP value of the table.

In Gene, a complex instance with 747 variables, the best algorithms are GALM, GABP, and LM. As in Cancer, the hybrid algorithms outperform the CEP of LM. The best values of SEP are obtained by GALM and LM. The low CEP of GABP contrasts with its high SEP value.

For the Soybean, another complex instance, only LM obtains competitive results in CEP. The rest of the algorithms get a CEP above 50%, what gives an idea about the relative complexity of the instance. With respect to the SEP, GA gets the highest value and BP, GABP, and GALM obtain similar SEP values. The lowest SEP value is reached by LM (also the clear winner with respect to the CEP).

Finally, in the Thyroid instance all algorithms except LM obtain always 7.28% of CEP. This is a bad result because in this data set 92.58% of the patterns belong to one class. That is, as in Diabetes with GA and GABP, the trained ANNs classify all the test patterns into the same class. The best value of SEP is obtained by LM. The worst one belongs to GA.

Table 6-4. Results of SEP in ANN training

SEP		BP	LM	GA	GABP	GALM
BC	Avg.	**0.49**	3.51	16.70	7.23	4.53
	Dev.	0.02	1.22	0.99	3.27	0.25
DI	Avg.	**14.47**	22.45	22.54	23.04	18.32
	Dev.	0.08	3.43	0.16	0.18	0.10
HE	Avg.	18.78	34.73	24.80	25.16	**14.96**
	Dev.	0.33	3.56	1.15	0.98	0.26
GE	Avg.	22.01	9.12	21.82	18.80	**8.52**
	Dev.	3.24	1.59	2.85	1.90	1.84
SO	Avg.	5.20	**2.20**	19.94	6.35	4.97
	Dev.	0.05	0.78	1.70	0.53	1.19
TH	Avg.	4.84	**1.25**	6.62	4.88	4.46
	Dev.	0.03	0.51	0.80	0.17	0.03

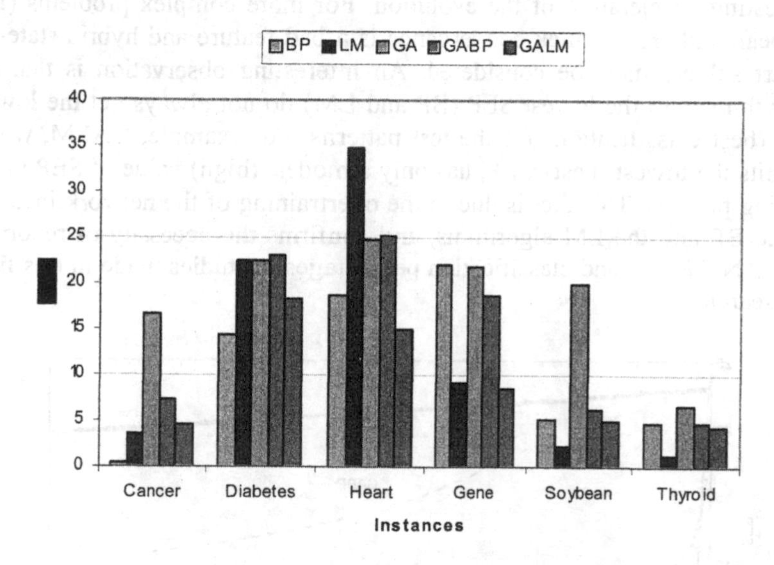

Figure 6-7. Comparison of the SEP among the algorithms.

In summary, focusing on CEP in pure algorithms, we observed that BP performed slightly more accurately than LM for the first three instances. In the rest, LM improves the results of BP. With respect to the hybrid algorithms, GALM is more accurate than GABP. Furthermore, it gets the best overall results in Cancer, Heart, and Gene and obtains competitive results in Diabetes and Soybean. GALM is the most accurate algorithm in three out of six instances and LM in two of them. This states the outstanding performance of LM. In the remaining instance (Diabetes), BP obtains the best CEP. We clearly deduce that the canonical GA is not a competitive algorithm for the ANN training problem but it can be improved in some cases by hybridizing it with other algorithms. With respect to SEP, the results hold, with the exception of Cancer, in which BP shows the best SEP.

We have traced the evolution of each algorithm for the Cancer instance to illustrate how the different algorithms work. In Figure 6-8 we plot the training SEP of the network in each step of the algorithm. For population-based algorithms (GA, GABP, and GALM) we trace the SEP value of the best fitness network. Each line represents the average SEP over 50 independent runs. We can observe that LM is the fastest algorithm to converge, followed by BP, they clearly stagnate before 100 epochs. This confirms the fast convergence of local search compared to GAs and hybrids, one of the main drawbacks of gradient search. The GA is the slowest algorithm. Its hybridization with BP, and especially with LM shows an interesting acceleration of the evolution. For more complex problems (like Soybean and Gene) quick convergence is a bad feature and hybrid state-of-the-art solvers must be considered. An interesting observation is that the algorithms with the lowest SEP (BP and LM) do not always get the lowest CEP (best classification) for the test patterns. For example, GALM, which exhibits the lowest (best) CEP, has only a modest (high) value of SEP in the training process. This fact is due to the overtraining of the network incurred by the BP and the LM algorithms, and confirms the necessity of reporting both, ANN errors and classification percentages in studies made in this field of research.

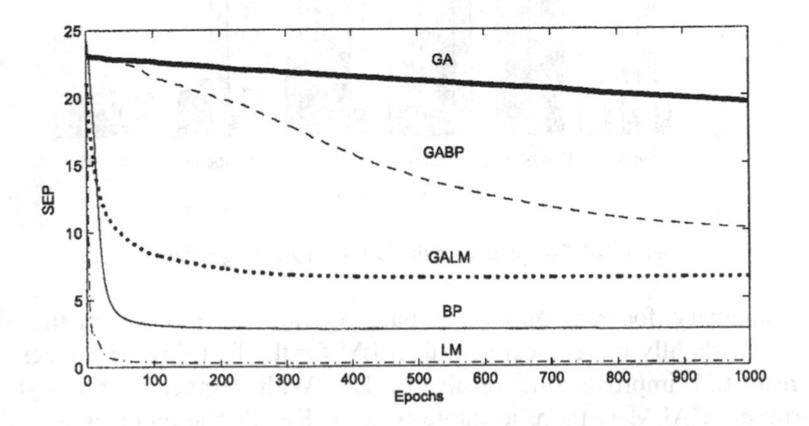

Figure 6-8. Average evolution of SEP for the algorithms on the Cancer instance.

5.2 Computational Effort

In this section we analyze the computational effort of the algorithms for solving each problem instance. In Table 6-5 we show the average and standard deviation of the execution times for 50 independent runs. In Figure 6-9 we display the average execution times.

We are using a set of different algorithms with different parameters, and hence we cannot draw globally valid conclusions about the relative computational effort of the algorithms for each problem. However, we consider an important issue in a technical work to report on efficiency as a guide for other researchers. A clear conclusion is that the numerical advantages of using somehow LM come at the price of higher computational times (Figure 6-9). Both LM and its hybrid GALM exhibited the highest computation times. The fastest algorithms in our experiments are GA and BP. This is a very interesting observation since many people think of GAs as very slow optimizers, what is not the case here. BP is expected to be fast since its time complexity is low; BP is both efficient and fast, maybe not the best in these two criteria, but its good performance in the two aspects explains its popularity.

Table 6-5. Execution time for the ANN training algorithms in seconds

Time (s)		BP	LM	GA	GABP	GALM
BC	Avg.	8.18	95.08	**2.40**	11.76	130.72
	Dev.	0.86	7.66	0.49	1.03	44.15
DI	Avg.	6.88	91.48	**2.20**	10.72	125.18
	Dev.	0.55	9.75	0.40	0.45	0.71
HE	Avg.	12.56	499.82	**4.84**	20.08	649.94
	Dev.	0.75	39.21	1.41	0.87	53.61
GE	Avg.	**8.36**	831.42	37.32	47.54	1164.34
	Dev.	0.48	4.13	0.47	2.57	250.47
SO	Avg.	**1.38**	748.18	15.22	16.90	926.94
	Dev.	0.49	1.23	0.41	0.64	242.34
TH	Avg.	**7.90**	274.64	29.48	39.14	409.18
	Dev.	0.30	1.55	0.54	1.37	41.48

We now present the relationship between execution time and test CEP to explicitly state the accuracy vs. time trade-off decision. In Figures 6-10 and 6-11 we plot the CEP values for the test set over execution time of the algorithms used in this work. We highlight with a shadow the algorithms that are not outperformed simultaneously in time and CEP by another algorithm. As we expected, the algorithms LM and GALM need a higher time than the others in all the instances because of the Jacobian computation of the error vector. However, one of them always gets the best CEP (except for Diabetes) and for this reason they appear highlighted in the figures. The GA is the fastest algorithm in the three first instances: those with a smaller number of variables (Cancer, Diabetes, and Heart). The hybrid algorithms are always slower than their pure components. The BP algorithm, despite its bad CEP values, appears always highlighted in the figures because of its low execution time. GA is highlighted in four of the six instances despite of its high CEP for the same reason as BP: its low execution time. In summary, the lowest CEPs are obtained always by LM, and GALM, and the lowest times always by BP and GA.

We are tempted to compare the "average" difficulty of solving the problems by noticing the relative position of its associated regions in Figures 6-10 and 6-11. However, we must be careful with this comparison because the algorithms have a different parameterization for each instance. The only algorithm maintaining the parameterization throughout the instances is GA, so for a "fair" comparison we take into account only GA. On the other hand, we must consider the number of classes of the instances. To explain this let us suppose that we have an ignorant classifier, that is, a classifier whose output is a random class for each input pattern. Using this classifier with an instance with two output classes we expect a CEP of 50% but if we use it in a three classes instance we expect a 66% of CEP. Hence, we need to normalize the CEP before comparing the difficulty of the instances.

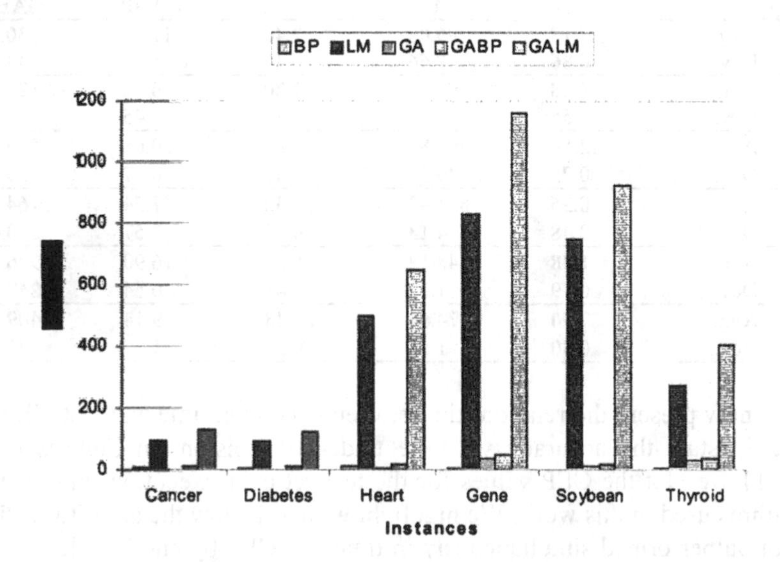

Figure 6-9. Average execution time for the ANN training algorithms.

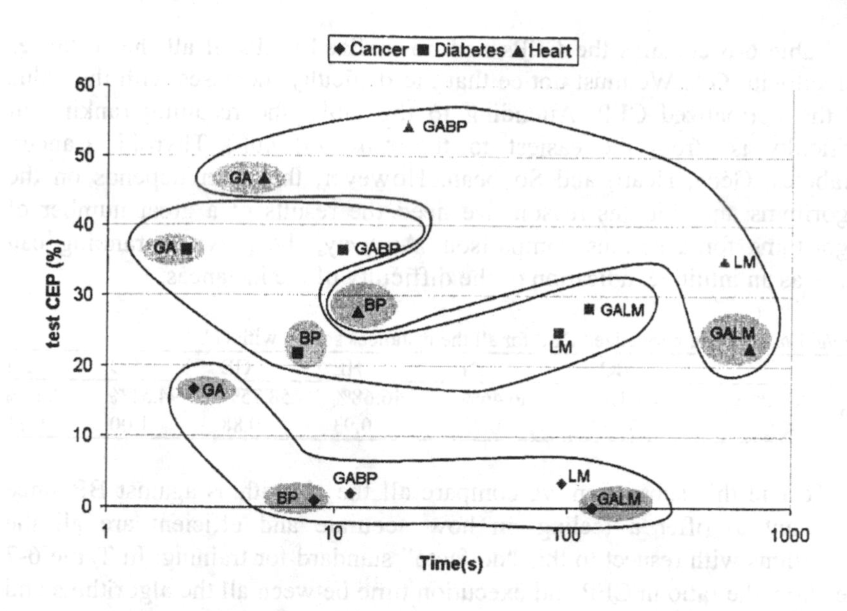

Figure 6-10. CEP against execution time of the algorithms in Cancer, Diabetes, and Heart.

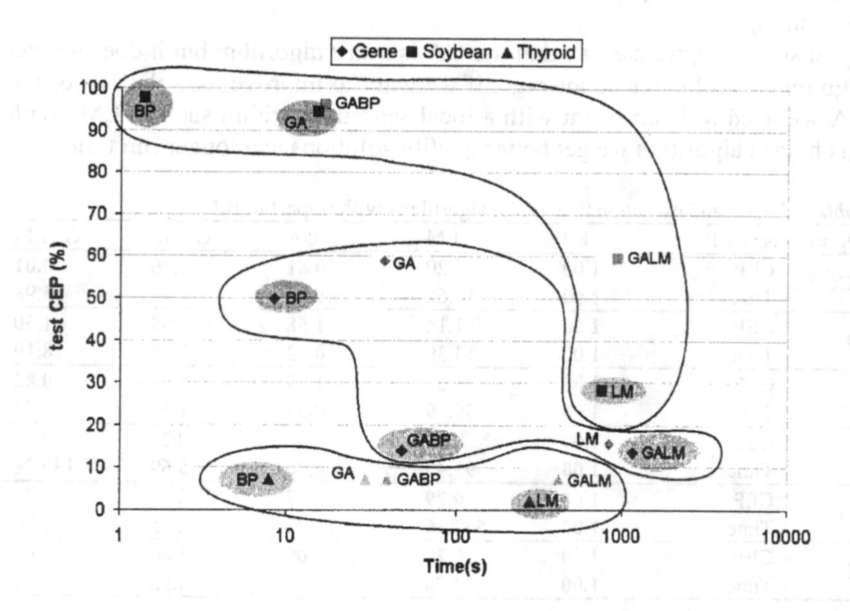

Figure 6-11. CEP against execution time of the algorithms in Gene, Soybean, and Thyroid.

Table 6-6 contains the CEPs and normalized CEPs of all the instances solved with GA. We must notice that the difficulty increases with the value of the normalized CEP. Attending to the table, the resulting ranking of difficulty is (from the easiest to the most difficult) Thyroid, Cancer, Diabetes, Gene, Heart, and Soybean. However, the order depends on the algorithms and, for this reason, we need the results of a great number of algorithms for a serious comparison. Anyway, the previous ranking can serve as an intuitive indication of the difficulty of the instances.

Table 6-6. CEP and normalized CEP for all the instances solved with GA

		BC	DI	HE	GE	SO	TH
GA	CEP	16.64%	36.46%	46.68%	58.75%	94.51%	7.28%
	nCEP	0.33	0.73	0.93	0.88	1.00	0.11

To end this subsection we compare all the algorithms against BP since we want to offer a feeling on how accurate and efficient are all the algorithms with respect to this “de facto” standard for training. In Table 6-7 we show the ratio in CEP and execution time between all the algorithms and BP. For the instances with a large number of variables (Gene, Soybean, and Thyroid) BP is the fastest algorithm. However, it never gets the lowest CEP. Depending on the application, the user must select his/her preferred method for training.

In summary, we can say that the GA is a fast algorithm but it does not get high quality solutions on average. If we want to improve the solutions of the GA we need to hybridize it with a local search algorithm such as LM. With this hybrid algorithm we get better quality solutions and longer run times.

Table 6-7. CEP and execution time of the algorithms with respect to BP

Algorithms vs BP		BP	LM	GA	GABP	GALM
BC	CEP	1.00	4.20	19.81	2.26	**0.01**
	Time	1.00	11.62	**0.29**	1.44	15.98
DI	CEP	**1.00**	1.14	1.68	1.68	1.30
	Time	1.00	13.30	**0.32**	1.56	18.19
HE	CEP	1.00	1.27	1.69	1.96	**0.82**
	Time	1.00	39.79	**0.39**	1.60	51.75
GE	CEP	1.00	0.31	1.18	0.28	**0.27**
	Time	**1.00**	99.45	4.46	5.69	139.28
SO	CEP	1.00	**0.29**	0.97	0.98	0.61
	Time	**1.00**	542.16	11.03	12.25	671.70
TH	CEP	1.00	**0.29**	1.00	1.00	1.00
	Time	**1.00**	34.76	3.73	4.95	51.79

5.3 Comparison Against Existing Approaches

There are many interesting works related to neural network training that also solve the instances tackled here. But, unfortunately, some of the results are not comparable with ours because they use a different definition of the training and test sets; this is why we consider as a capital issue to adhere to a standard way of evaluations like the one proposed by Prechelt (1994). However, we did find some works for meaningful comparisons (Table 6-8).

For the Cancer instance we found that the best mean CEP in Land and Albertelli (1998) is 2%. This work uses 524 patterns for the training set and the rest for the test set, that is, almost exactly our configuration with only one pattern changed (a minor detail), and therefore the results can be compared. The same pattern configuration is present in the work of Yao and Liu (1997), where their EPNet algorithm gets neural networks reaching 1.4% of CEP. Finally, in Islam and Yao (2003) we find a CEP of 1.2% while it is reduced down to 1.1% in Ragg et al. (1997). All these results are outperformed by our GALM hybrid with a 0.01% CEP, the best so far solution to our knowledge.

In Diabetes, a CEP value of 30.11% is reached in Erhard et al. (1998) with the same network architecture as here. In Yao and Liu (1997) the authors report a still better 22.4% of CEP for this problem. Our BP, LM, and GALM outperform the first work, while the second one is outperformed by our BP (21.75%). The best CEP we have found in the literature for Diabetes is 19.60% in Islam and Yao (2003).

In Erhard et al. (1998) the authors reported a 45.71% of CEP for the Heart instance using the same architecture. Many of our algorithms outperform this CEP measure (best: GALM 22.70%).

In Islam and Yao (2003), using a constructive algorithm, a CEP of 7.60% is obtained for the Soybean instance: the best CEP for this instance to our knowledge. Our best result for this problem is found with LM (28.11%).

Finally, for Thyroid a CEP of 1% is achieved in Ragg et al. (1997) slightly better than our best algorithm for this instance LM with a 2.14% of CEP.

Table 6-8. Comparison among CEP obtained in this work and the best found in the literature

CEP(%)	BP	LM	GA	GABP	GALM	Literature
BC	0.84	3.53	16.64	1.90	**0.01**	1.10
DI	21.75	24.75	36.46	36.46	28.21	**19.60**
HE	27.67	35.13	46.68	54.22	**22.70**	45.71
GE	49.76	15.51	58.75	13.89	**13.51**	-
SO	97.74	28.11	94.51	96.12	59.40	**7.60**
TH	7.28	2.14	7.28	7.28	7.28	**1.00**

6. CONCLUSIONS

In this chapter we have tackled the neural network training problem with genetic algorithms. We have discussed some key issues such as representation of the networks, fitness function, and hybridization. To illustrate the technique we have solved six classification problems from the domain of Bioinformatics by training a classifier neural network. The problems are: the diagnosis of breast cancer, the diagnosis of diabetes in the Pima Indians, the diagnosis of heart disease, the splice junction detection, the recognition of soybean diseases, and the diagnosis of thyroid hyper- or hypofunction. As an initial remark, we must emphasize the importance of standard comparisons. It is very important the use of standard rules to preprocess the pattern set in order to compare the results against other authors following the same preprocessing rules. This benefits the comparison between different training algorithms.

We have compared the results of the GA, two classical algorithms, and two hybrid algorithms between them (GABP and GALM). Our results show that the problem-specific algorithms (BP and LM) get lower classification error than the other more general search procedures only in some cases. A clear conclusion is that LM must be seriously considered in the metaheuristics field. With respect to the hybrids, the algorithm GALM outperforms in three of the six instances the classification error of the problem-specific algorithms. This makes GALM to look as a promising algorithm for neural network training.

ACKNOWLEDGEMENTS

This work has been partially funded by the Ministry of Science and Technology (MCYT) and Regional Development European Found (FEDER) under contract TIC2002-04498-C05-02 (the TRACER project, http://tracer.lcc.uma.es). Francisco Chicano is supported by grant (BOJA 68/2003) from the Junta de Andalucía (Spain).

APPENDIX A: BACKPROPAGATION

The Backpropagation (BP) algorithm (Rumelhart et al., 1986) is a classical domain-dependent technique very used in supervised training. It works by measuring the output error, calculating the gradient of this error, and adjusting the ANN weights (and biases) in the descending gradient direction. Hence, BP is a gradient-descent local search procedure: this means a probably high accuracy but also a foreseeable stagnation in local optima in complex search landscapes.

In order to apply BP, first, we define the squared error of the ANN for a set of patterns as

$$E = \sum_{p=1}^{P} \sum_{i=1}^{S} (t_i^p - o_i^p)^2 \tag{5}$$

The actual value of the previous expression depends on the weights of the network. The basic BP algorithm calculates the gradient of E and updates the weights by moving them in the gradient-descent direction. This can be summarized with the expression

$$w_{ij}(t+1) = w_{ij}(t) - \eta \frac{\partial E}{\partial w_{ij}} \tag{6}$$

where the parameter $\eta > 0$ is the learning rate that controls the learning speed. A more general BP algorithm adds a momentum term to the previous expression in order to increase the stability of the search process. Then, the final expression for the BP algorithm is

$$w_{ij}(t+1) = w_{ij}(t) + \alpha \Delta w_{ij}(t) - \eta \frac{\partial E}{\partial w_{ij}} \tag{7}$$

where $\Delta w_{ij}(t)$ is the change in the weight w_{ij} at step t, and α is the momentum constant, whose value is in the interval [0,1). With this term, the algorithm accelerates the minimization of the error when the error function does not change (in smooth zones of the function). The pseudocode of the BP algorithm is shown in Figure 6-12.

```
InitializeWeights;
for all i, j do
        Δwij := 0;
endfor;
while not StopCriterion do
        for all i,j do
                Δwij := αΔwij - η∂E/∂wij;
                wij := wij + Δwij;
        endfor;
endwhile;
```

Figure 6-12. Pseudocode of the BP algorithm.

APPENDIX B: LEVENBERG-MARQUARDT

The Levenberg-Marquardt (LM) algorithm (Hagan and Menhaj, 1994) is an approximation to the Newton method used also for training ANNs. This Newton method approximates the error of the network with a second order expression, which contrasts to the first order information used in the Backpropagation algorithm. LM is very popular in the ANN domain. Curiously, it is not so popular in the metaheuristics field. LM updates the ANN weights following the expression

$$\Delta \mathbf{w} = -\left[\mu \mathbf{I} + \sum_{p=1}^{P} J^{p}(\mathbf{w})^{T} J^{p}(\mathbf{w})\right]^{-1} \nabla E(\mathbf{w}) \quad (8)$$

where $J^p(\mathbf{w})$ is the Jacobian matrix of the vector $\mathbf{e}^p(\mathbf{w})$ evaluated in $\mathbf{w}$, and $\mathbf{I}$ is the identity matrix. The vector $\mathbf{e}^p(\mathbf{w})$ is the error of the network for pattern p, that is, $\mathbf{e}^p(\mathbf{w}) = \mathbf{t}^p - \mathbf{o}^p(\mathbf{w})$. The parameter μ changes during the training. It can be multiplied or divided by the factor β. The exact steps of the LM algorithm are depicted in Figure 6-13. In the outer loop it calculates the network output, the error vectors, and the Jacobian matrix for each pattern. Then, it enters the inner loop, where it computes $\Delta\mathbf{w}$ using Eq. (8) and recalculates the error with $\mathbf{w}+\Delta\mathbf{w}$ as network weights. If the error does not decrease it multiplies μ by β and goes to the beginning of the inner loop, where it calculates $\Delta\mathbf{w}$ with the new value of μ. If the error decreases, then it leaves the inner loop, divides μ by β, and adds $\Delta\mathbf{w}$ to the network weight vector. Then, it goes back to the beginning of the outer loop and iterates again until the stop criterion is fulfilled.

```
InitializeWeights;
while not StopCriterion do
      calculate e^p(w);
      e1 := Σ_{p=1}^{P} [e^p(w)]^2;   // error
      calculate J^p(w);
      repeat
            calculate Δw;
            e2 := Σ_{p=1}^{P} [e^p(w+Δw)]^2;
            if (e1 <= e2) then
                  μ := μ * β;
            endif;
      until (e2 < e1);
      μ := μ / β;
      w := w+Δw;
endwhile;
```

Figure 6-13. Pseudocode of the LM algorithm.

REFERENCES

Alander, J. T., 1994, Indexed bibliography of genetic algorithms and neural networks. Technical Report 94-1-NN, University of Vaasa, Department of Information Technology and Production Economics.

Alba, E., 1993, Aplicación de los algoritmos genéticos para el diseño de redes neuronales, *Informática y Automática* **26**(2):22-35 (text in spanish).

Alba, E., Aldana, J. F., and Troya, J. M., 1993, Full automatic ANN design: a genetic approach, in: *New Trends in Neural Computation, International Workshop on artificial Neural Networks*, J. Mira, J. Cabestany, A., and Prieto, eds., Springer-Verlag, Sitges, Spain, pp. 399-404.

Bäck, T., 1996, *Evolutionary Algorithms in Theory and Practice: Evolution Strategies, Evolutionary Programming, Genetic Algorithms*, Oxford University Press, New York.

Bäck, T., Fogel, D. B., Whitley, D., and Angeline, P. J., 2000, Mutation operators, in: *Evolutionary Computation 1. Basic Algorithms and Operators*, T. Bäck, D. B. Fogel, and T. Michalewicz, eds., IOP Publishing Lt, pp. 237-255.

Bennett, K. P., and Mangasarian, O. L., 1992, Robust linear programming discrimination of two linearly inseparable sets, *Optimization Methods and Software.* **1**:23-34.

Blum, C., and Roli, A., 2003, Metaheuristics in combinatorial optimization: overview and conceptual comparison, *ACM Computing Surveys.* **35**(3):268-308.

Booker, L. B., Fogel, D. B., Whitley, D., Angeline, P. J., and Eiben, A. E., 2000, Recombination, in: *Evolutionary Computation 1. Basic Algorithms and Operators*, Bäck, T., Fogel, D. B., and Michalewicz, T., eds., IOP Publishing Lt, pp. 256-307.

Cantú-Paz, E., 2003, Pruning neural networks with distribution estimation algorithms, in: *Proceedings of GECCO 2003 (LNCS 2723)*, E. Cantú-Paz et al., eds., Springer-Verlag, Chicago, USA, pp. 790-800.

Cao, C., Wang, X., and Lu, M.-P., 2003, Direct torque control based on FNN and optimization, in: *Proceedings of the Second International Conference on Machine Learning and Cybernetics*, IEEE Computer Society Press, Xian, China, pp. 760-764.

Cheng, W.-S., 2004, An application of adaptive-genetic neural algorithm to sinter's BTP process, in: *Proceedings of the Third International Conference on Machine Learning and Cybernetics*, IEEE Computer Society Press, Shanghai, China, pp. 3356-3360.

Cotta, C., Alba, E., Sagarna, R., and Larrañaga, P., 2001, Adjusting weights in artificial neural networks using evolutionary algorithms, in: *Estimation of Distribution Algorithms. A New Tool for Evolutionary Computation*, P. Larrañaga, and J. A. Lozano, eds., Kluwer Academic Publishers, Boston, USA, pp. 357-373.

Cotta, C., and Troya, J. M., 1998, On decision-making in strong hybrid evolutionary algorithms, in: *Tasks and Methods in Applied Artificial Intelligence*, Lecture Notes in Artificial Intelligence, vol. 1415, Del Pobil, A. P., Mira, J., and Ali, M., eds., Springer-Verlag, Berlin, Germany, 418-427.

Davis, L., 1991, *Handbook of Genetic Algorithms*, Van Nostrand Reinhold, New York.

Detrano, R., Janosi, A., Steinbrunn, W., Pfisterer, M., Schmid, J., Sandhu, S., Guppy, K., Lee, S., and Froelicher, V., 1989, International application of a new probability algorithm for the diagnosis of coronary artery disease, *American Journal of Cardiology* **64**:304-310.

Erhard, W., Fink, T., Gutzmann, M. M., Rahn, C., Doering, A., and Galicki, M., 1998, The improvement and comparison of different algorithms for optimizing neural networks on the MasPar MP-2, in: *Neural Computation - NC'98*, M. Heiss, ed., ICSC Academic Press, Vienna, Austria, pp. 617-623.

Eshelman, L. J., 2000, Genetic algorithms, in: *Evolutionary Computation 1. Basic Algorithms and Operators*, T. Bäck, D. B. Fogel, and T. Michalewicz, eds., IOP, pp. 64-80.

Gengyin, L., Ming, Z., and Zhiyuan, Z., 2003, Research on power quality disturbance automatic recognition and location, in: *Proceedings of the Power Engineering Society General Meeting,* IEEE Computer Society Press, Toronto, Canada, pp. 687-691.

Gennari, J. H., Langley, P., and Fisher, D., 1989, Models of incremental concept formation, *Artificial Intelligence* **40**:11-61.

Gilson, M., Py, J. S., Brault, J. J., and Sawan, M., 2003, Training recurrent pulsed networks by genetic and taboo methods, in: *Proceedings of the Canadian Conference on Electrical and Computer Engineering*, IEEE Computer Society Press, Montreal, CA, pp. 1857-1860.

Hagan, M. T., and Menhaj, M. B., 1994, Training feedforward networks with the Marquardt algorithm, *IEEE Transactions on Neural Networks* **5**(6):989-993.

Holland, J. H., 1975, *Adaptation in Natural and Artificial Systems*, the University of Michigan Press, Ann Arbor, Michigan.

Islam, M. M., and Yao, X., 2003, A constructive algorithm for training cooperative neural network ensembles, *IEEE Transactions on Neural Networks* **14**(4):820-834.

Ku, K. W. C., Mak, M. W., and Siu, W. C., 1999, Adding learning to cellular genetic algorithms for training recurrent neural networks, *IEEE Transactions on Neural Networks* **10**(2):239-252.

Kuri-Morales, A. F., Ortiz-Posadas, M. R., Zenteno, D., and Peñaloza, R. P., 2003, Classification of sperm cells according to their chromosomic content using a neural network trained with a genetic algorithm, in: *Proceedings of the 25th Annual Intemational Conference of the IEEE EMBS,* IEEE Computer Society Press, Cancun, Mexico, pp. 2253-2256.

Land, W. H., and Albertelli, L. E., 1998, Breast cancer screening using evolved neural networks, in: *Proceedings of the IEEE International Conference on Systems, Man, and Cybernetics*, IEEE Computer Society Press, San Diego, USA, pp. 1619-1624.

Leung, F. H. F., Lam, H. K., Ling, S. H., and Tam, P. K. S., 2003a, Tuning of the structure and parameters of a neural network using an improved genetic algorithm, *IEEE Transactions on Neural Networks* **14**(1):79-88.

Leung, K. F., Leung, F. H. F., Lam, H. K., and Ling, S. H., 2004, On interpretation of graffiti digits and characters for ebooks: Neural-fuzzy network and genetic algorithm approach. *IEEE Transactions on Industrial Electronics* **51**(2):464-471.

Leung, K., Leung, F., Lam, H., and Tam, P., 2003b, Neural fuzzy network and genetic algorithm approach for cantonese speech command recognition, in: *Proceedings of the 12th IEEE International Conference on Fuzzy Systems,* IEEE Computer Society Press, St. Louis, USA, pp. 208-213.

Ling, S. H., Leung, F. H. F., Lam, H. K., Lee, Y.-S., and Tam, P. K. S., 2003, A novel genetic-algorithm-based neural network for short-term load forecasting, *IEEE Transactions on Industrial Electronics* **50**(4):793-799.

Lisboa, P. J. G., Vellido, A., and Edisbury, B., 2000, *Business Applications of Neural Networks. The State-of-the-Art of Real-World Applications*, World Scientific.

Mangasarian, O. L., Setiono, R., and Wolberg, W. H., 1990, Pattern recognition via linear programming: theory and application to medical diagnosis, in: *Large-Scale Numerical Optimization*, T. F. Coleman, and Y. Li, eds., SIAM Publications, Philadelphia, USA, pp. 22-31.

McClelland, J. L., and Rumelhart, D. E., 1986, *Parallel Distributed Processing: Explorations in the Microstructure of Cognition.* The MIT Press, Cambridge, USA.

Moon, S.-W., and Kong, S.-G., 2001, Block-based neural networks, *IEEE Transactions on Neural Networks* **12**(2):307-317.

Noordewier, M. O., Towell, G. G., and Shavlik, J. B., 1991, Training knowledge-based neural networks to recognize genes in DNA, in: *Advances in Neural Information Processing Systems,* R. P. Lippmann, J. E. Moody, and D. S. Touretzky, eds., Morgan Kaufmann Publishers Inc., Denver, USA, pp. 530-536.

Parker, G. B., and Lee, Z., 2003, Evolving neural networks for hexapod leg controllers, in: *Proceedings of the 2003 IEEE/RSJ International Conference on Intelligent Robots and Systems,* IEEE Computer Society Press, Las Vegas, USA, pp. 1376-1381.

Pattnaik, S. S., Khuntia, B., Panda, D. C., Neog, D. K., and Dutta, S. D. M., 2005, Application of a genetic algorithm in an artificial neural network to calculate the resonant frequency of a tunable single-shorting-post rectangular-patch antenna, *International Journal of RF and Microwave Computer-Aided Engineering* **15**(1):150-144.

Prechelt, L., 1994, Proben1—A set of neural network benchmark problems and benchmarking rules. Technical Report 21, Fakultät für Informatik Universität Karlsruhe, GE, 76128.

Ragg, T., Gutjahr, S., and Sa, H., 1997, Automatic determination of optimal network topologies based on information theory and evolution, in: *Proceedings of the 23rd EUROMICRO Conference,* IEEE Computer Society Press, Budapest, pp. 549-555.

Rosenblatt, F., 1962, *Principles of Neurodynamics*, Spartan Books, New York.

Rumelhart, D., Hinton, G., and Williams, R., 1986, Learning representations by backpropagation errors, *Nature* **323**:533-536.

Schiffmann, W., Joost, M., and Werner, R., 1992, Optimization of the backpropagation algorithm for training multilayer perceptrons. Technical report, Institute of Physics, University of Koblenz, Koblenz, Germany.

Smith, J. W., Everhart, J. E., Dickson, W. C., Knowler, W. C., and Johannes, R. S., 1988, Using the ADAP learning algorithm to forecast the onset of diabetes mellitus, in: *Proceedings of the Twelfth Symposium on Computer Applications in Medical Care,* IEEE Computer Society Press, Washington D.C., USA, pp. 261-265.

Talbi, E.-G., 2002, A taxonomy of hybrid metaheuristics, *Journal of Heuristics* **8**(2):541-564.

Tan, M., and Eshelman, L., 1988, Using weighted networks to represent classification knowledge in noisy domains, in: *Proceedings of the 5th International Conference on Machine Learning,* Ann Arbor, USA, pp. 121-134.

Tarca, L. A., Grandjean, B. P. A., and Larachi, F., 2004, Embedding monotonicity and concavity in the training of neural networks by means of genetic algorithms application to multiphase flow, *Computers and Chemical Engineering* **28**(9):1701-1713.

Wang, K., Gelgele, H. L., Wang, Y., Yuan, Q., and Fang, M., 2003, A hybrid intelligent method for modelling the EDM process, *International Journal of Machine Tools Manufacture* **43**(10):995-999.

Wolberg, W. H., 1990, Cancer diagnosis via linear programming, *SIAM News* **23**(5):1–18.

Wolberg, W. H., and Mangasarian, O. L., 1990, Multisurface method of pattern separation for medical diagnosis applied to breast cytology, *Proceedings of the National Academy of Sciences* **87**(23):9193-9196.

Xiaoyun, S., Donghui, L., Kai, Z., Liweil, G., Ran, Z., and Jianye, L., 2004, Neural network with adaptive genetic algorithm for eddy current nondestructive testing, in: *Proceedings of the 5th World Congress on Intelligent Control and Automation,* IEEE Computer Society Press, Hangzhou, China, pp. 2034-2037.

Yao, X., 1999, Evolving artificial neural networks, *Proceedings of the IEEE* **87**(9):1423-1447.

Yao, X., and Liu, Y., 1997, A new evolutionary system for evolving artificial neural networks, *IEEE Transactions on Neural Networks* **8**:694-713.

Chapter 7

SCATTER SEARCH

Manuel Laguna[1] and Rafael Martí[2]
[1]*Leeds School of Business, University of Colorado at Boulde, USA;*
[2]*Dpto. de Estadística e Investigación Operativa, Universidad de Valencia, Spain*

Abstract: This chapter discusses the principles and foundations behind scatter search and its application to the problem of training neural networks. Scatter search is an evolutionary method that has been successfully applied to a wide array of hard optimization problems. Scatter search constructs new trial solutions by combining so-called reference solutions and employing strategic designs that exploit context knowledge. In contrast to other evolutionary methods like genetic algorithms, scatter search is founded on the premise that systematic designs and methods for creating new solutions afford significant benefits beyond those derived from recourse to randomization. Our implementation goal is to create a combination of the five elements in the scatter search methodology that proves effective when searching for optimal weight values in a multilayer neural network. Through experimentation, we show that our instantiation of scatter search can compete with the best-known training algorithms in terms of training quality while keeping the computational effort at a reasonable level.

Key words: Metaheuristics; neural networks; optimization.

1. INTRODUCTION

Scatter search (SS) was first introduced by Glover (1977) as a heuristic for integer programming. In the original proposal, solutions are purposely (i.e., non-randomly) generated to take account of characteristics in various parts of the solution space. The orientation of SS is to explore systematically relative to a set of reference points that typically consists of good solutions obtained by prior problem solving efforts, where the criteria for "good" are not restricted to objective function values. In this way, SS shares with other evolutionary methods the philosophy of operating on a set of solutions rather than on a single solution at a time. It also shares the conceptual framework of embedding procedures for combining these solutions to create new ones. However, the meaning of "combining" and the way is carried out have a rather special origin and character in the SS setting. A distinguishing feature of scatter search is its intimate association with the Tabu Search (TS) metaheuristic (Glover and Laguna 1997), which explains its adoption of the principle that search can benefit by incorporating special forms of adaptive memory, along with procedures particularly designed for exploiting that memory. In fact, scatter search and tabu search share common origins, and initially SS was simply considered one of the component processes available within the TS framework. However, most of the TS literature and the preponderance of TS implementations have disregarded this component, with the result that the merits of scatter search did not come to be recognized until quite recently. Nowadays it is a well established methodology within the metaheuristic community, although general awareness of the procedure still lags behind that of other population-based methods such as genetic algorithms and evolutionary strategies.

Artificial neural networks (ANNs) offer a general framework for representing non-linear mappings from several input variables to several output variables. They are built by tuning a set of parameters known as weights, and can be considered as an extension of the multitude of conventional mapping techniques discussed in the literature. While in classification or recognition problems the outputs of a neural network are categories, in prediction or approximation problems they are continuous variables. Although this paper is focused on the prediction problem, most of the key issues in the functionality of a neural network are common to both.

In this chapter we have considered the most commonly employed architecture for prediction and classification: *a multilayer feed-forward network with a single hidden layer*. A schematic representation of such a network appears in Figure 7-1.

Note that the weights in the network depicted in Figure 7-1 are numbered sequentially starting with the first input to the first hidden neuron. Therefore, the weights for all the inputs to the first hidden neuron are w_1 to w_n. The bias term for the first hidden neuron is w_{n+1}.

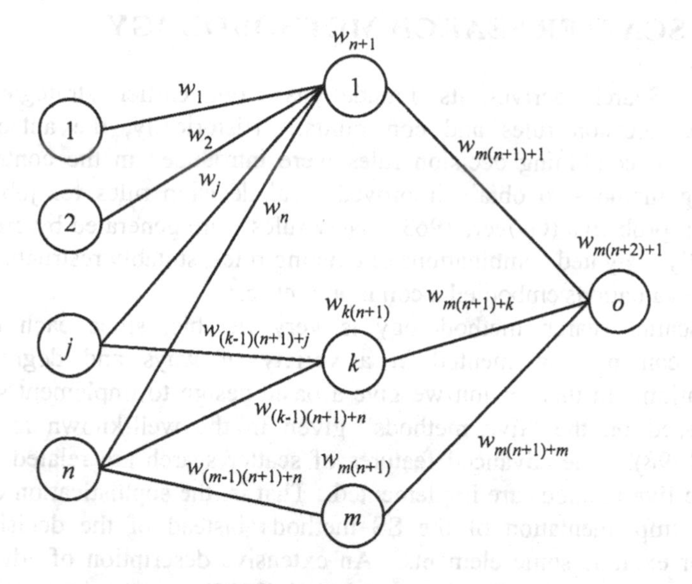

Figure 7-2. Neural network with one hidden layer and one output.

In the process of training the neural network (*supervised learning*), the problem is to find the values of the weights w that minimize the error across a set of input/output pairs (patterns). This set is referred to as the training set E. For a single output and input vector x, the error measure is typically the root mean squared difference (RMSE) between the predicted output $p(x,w)$ and the actual output value $f(x)$ for all the elements x in E; therefore, the training is an unconstrained nonlinear optimization problem, where the decision variables are the weights and the objective is to reduce the training error:

$$\underset{w}{Min}\ error(E,w) = \sqrt{\frac{\sum_{x \in E} (f(x) - p(x,w))^2}{|E|}}$$

In the remainder of the chapter, we first introduce in Section 2 the scatter search methodology, and then describe in Section 3 the SS method by Laguna and Martí (2000) for the neural network training. Section 4 is devoted to the computational experiments and the chapter finishes with relevant conclusions.

2. SCATTER SEARCH METHODOLOGY

Scatter Search derives its foundations from earlier strategies for combining decision rules and constraints. Historically, the antecedent strategies for combining decision rules were introduced in the context of scheduling methods to obtain improved local decision rules for job shop scheduling problems (Glover, 1963). New rules were generated by creating numerically weighted combinations of existing rules, suitably restructured so that their evaluations embodied a common metric.

The scatter search methodology is very flexible, since each of its elements can be implemented in a variety of ways and degrees of sophistication. In this section we give a basic design to implement scatter search based on the "five methods" given in the well-known *template* (Glover 1998). The advanced features of scatter search are related to the way these five methods are implemented. That is, the sophistication comes from the implementation of the SS methods instead of the decision to include or exclude some elements. An extensive description of advanced designs can be found in Laguna and Martí (2003) and the integration of additional elements with tabu search memory is treated in Glover (2004).

The fact that the mechanisms within scatter search are not restricted to a single uniform design allows the exploration of strategic possibilities that may prove effective in a particular implementation. These observations and principles lead to the following "five-method template" for implementing scatter search.

1. A *Diversification Generation Method* to generate a collection of diverse trial solutions, using an arbitrary trial solution (or seed solution) as an input.
2. An *Improvement Method* to transform a trial solution into one or more enhanced trial solutions. (Neither the input nor the output solutions are required to be feasible, though the output solutions will more usually be expected to be so. If no improvement of the input trial solution results, the "enhanced" solution is considered to be the same as the input solution.)
3. A *Reference Set Update Method* to build and maintain a reference set consisting of the b "best" solutions found (where the value of b is typically small, e.g., no more than 20), organized to provide efficient accessing by other parts of the method. Solutions gain membership to the reference set according to their quality or their diversity.
4. A *Subset Generation Method* to operate on the reference set, to produce a subset of its solutions as a basis for creating combined solutions.

5. A *Solution Combination Method* to transform a given subset of solutions produced by the Subset Generation Method into one or more combined solution vectors.

The Diversification Generation Method is used to build a large set P of diverse solutions. The size of P (*PSize*) is typically at least 10 times the size of *RefSet*. The initial reference set is built according to the Reference Set Update Method. It could consist of selecting b distinct and maximally diverse solutions from P.

The reference set, *RefSet*, is a collection of both high quality solutions and diverse solutions that are used to generate new solutions by way of applying the Combination Method. We can use a simple mechanism to construct an initial reference set and then update it during the search. The size of the reference set is denoted by $b = b_1 + b_2 = |RefSet|$. The construction of the initial reference set starts with the selection of the best b_1 solutions from P. These solutions are added to *RefSet* and deleted from P. For each solution in *P-RefSet*, the minimum of the distances to the solutions in *RefSet* is computed. Then, the solution with the maximum of these minimum distances is selected. This solution is added to *RefSet* and deleted from P and the minimum distances are updated. The process is repeated b_2 times, where $b_2 = b - b_1$. The resulting reference set has b_1 high quality solutions and b_2 diverse solutions.

Solutions in *RefSet* are ordered according to quality, where the best solution is the first one in the list. The simplest form of the Subset Generation Method consists of generating all pairs of reference solutions. That is, the method would focus on subsets of size 2 resulting in $(b^2\text{-}b)/2$ of them when all solutions in the reference set are new (that is, they have not been combined). After the first iteration, the number of subsets generated depends on the number of new solutions admitted to the *RefSet*. The subsets are put in a list and then are selected one at a time in lexicographical order to apply the Solution Combination Method that generates one or more trial solutions. These trial solutions are subjected to the Improvement Method.

The Reference Set Update Method is applied once again. A typical and quite simple form of the application of this method is the so-called *static update*. Trial solutions that are constructed as combination of reference solutions are placed in a solution pool, denoted by *Pool*. After the application of both the Combination Method and the Improvement Method, the *Pool* is full and the reference set is updated. The new reference set consists of the best b solutions from the solutions in the current reference set and the solutions in the pool, i.e., the update reference set contains the best b solutions in $RefSet \cup Pool$.

If *RefSet* changes after the application of the Reference Set Update Method a flag indicates that at least one new solution has been inserted in the reference set. The procedure terminates after all subsets generated within the current iteration are subjected to the Combination Method and none of the improved trial solutions are admitted to *RefSet* under the rules of the Reference Set Update Method. Figure 7-2 shows a diagram to illustrate a basic design of the method.

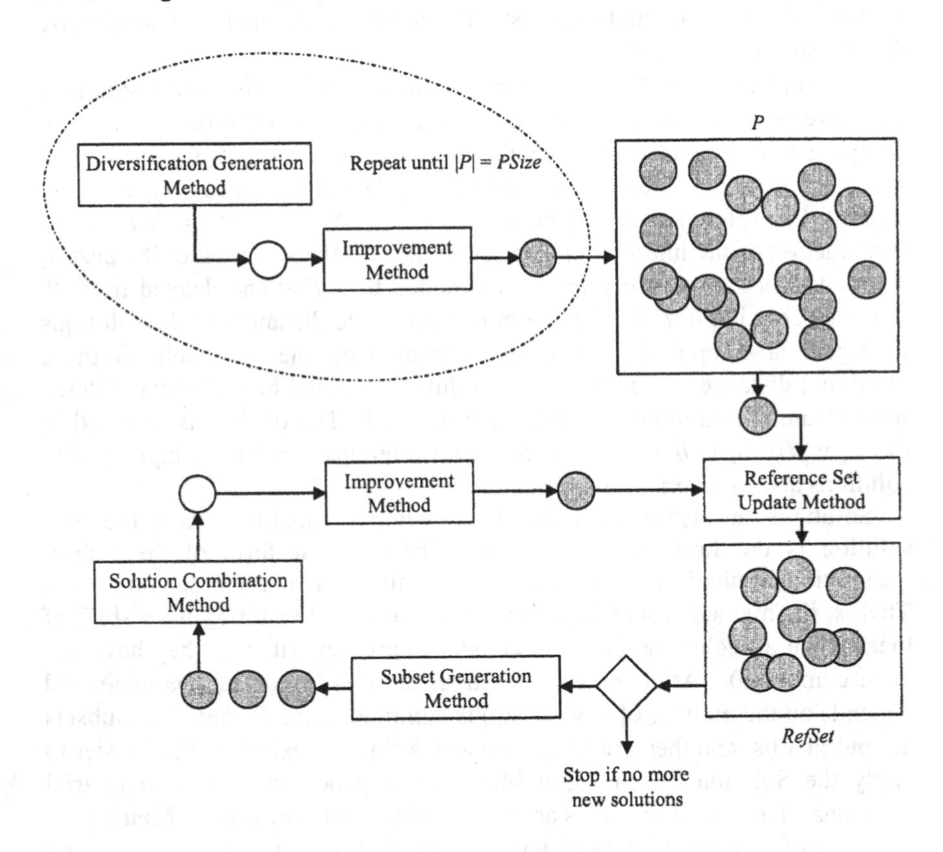

Figure 7-2. Basic scatter search design.

2.1 Advanced Designs

In the basic design, the new solutions that become members of *RefSet* are not combined until all eligible pairs are subjected to the Combination Method. The new reference set is built with the best solutions in the union of *Pool* and the solutions currently in *RefSet*. The alternative to this static update is the **Dynamic Update** strategy, which applies the Combination

Method to new solutions in a manner that combines new solutions faster than in the basic design. That is, if a new solution is admitted to the reference set, the goal is to allow this new solution to be subjected to the Combination Method as quickly as possible. Instead of waiting until all the combinations have been performed to update the reference set, if a new trial solution warrants admission in the reference set, the set is immediately updated before the next combination is performed. Therefore, there is no need for an intermediate pool in this design, since solutions are either discarded or become part of the *RefSet* as soon as they are generated.

RefSet Rebuilding is another advanced design that is triggered when no new trial solutions are admitted to the reference set. This update adds a mechanism to partially rebuild the reference set when the Combination and Improvement Methods do not provide solutions of sufficient quality to displace current reference solutions. The *RefSet* is partially rebuilt with a diversification update that works as follows and assumes that the size of the reference set is $b = b_1 + b_2$. Solutions $x^{b_1+1},\ldots,x^b$ are deleted from *RefSet*. The Diversification Generation Method is reinitialized considering that the goal is to generate solutions that are diverse with respect to the reference solutions $x^1,\ldots,x^{b_1}$. Then, the Diversification Generation Method is used to construct a set P of new solutions. The b_2 solutions $x^{b_1+1},\ldots,x^b$ in *RefSet* are sequentially selected from P with the criterion of maximizing the diversity. It is usually implemented with a distance measure defined in the context of the problem being solved. Then, maximum diversity is achieved by maximizing the minimum distance. The max-min criterion, which is part of the Reference Set Update Method, is applied with respect to solutions $x^1,\ldots,x^{b_1}$ when selecting solution x^{b_1+1}, then it is applied with respect to solutions $x^1,\ldots,x^{b_1+1}$ when selecting solution x^{b_1+2}, and so on.

Solution Combination Methods in scatter search typically are not limited to combining just two solutions and therefore the **Subset Generation Method** in its more general form consists of creating subsets of different sizes. The scatter search methodology assures that the set of combined solutions may be produced in its entirety at the point where the subsets of reference solutions are created. Therefore, once a given subset is created, there is no merit in creating it again. This creates a situation that differs noticeably from those considered in the context of genetic algorithms, where the combinations are typically determined by the spin of a roulette wheel.

The procedure for generating subsets of reference solutions uses a strategy to expand pairs into subsets of larger size while controlling the total number of subsets to be generated. In other words, the mechanism avoids the extreme type of process that creates all the subsets of size 2, then all the subsets of size 3, and so on until reaching the subsets of size b-1 and finally the entire *RefSet*. This approach clearly would not be practical, considering that there are 1013 subsets in a reference set of a typical size $b = 10$. Even for a smaller reference set, combining all possible subsets is not effective because many subsets will be almost identical. The following approach selects representative subsets of different sizes by creating subset types:

- Subset Type 1: all 2-element subsets.
- Subset Type 2: 3-element subsets derived from the 2-element subsets by augmenting each 2-element subset to include the best solution not in this subset.
- Subset Type 3: 4-element subsets derived from the 3-element subsets by augmenting each 3-element subset to include the best solutions not in this subset.
- Subset Type 4: the subsets consisting of the best i elements, for $i = 5$ to b.

We point out that an earlier proposal for generating subsets of different sizes using tabu search mechanisms, which apparently remains unexplored, appears in Glover (1994).

3. THE TRAINING METHOD

In this section we describe Laguna and Martí (2000) adaptation of scatter search to the neural network training problem. A pseudo-code of the algorithm is outlined in Figure 7-3 in which we refer to w as a solution to the training problem.

The procedure starts with the input and output data normalization. After this normalization, an initial reference set (*RefSet*) of b solutions is created. A set P of *PSize* solutions w (bounded between *wlow* and *whigh*) is built with the following diversification method, based on a controlled randomization scheme (Glover, Laguna and Martí, 2000). The range (*wlow, whigh*) is subdivided into 4 sub-ranges of equal size. Then, a solution w is constructed in two steps. First a sub-range is randomly selected. The probability of selecting a sub-range is inversely proportional to its frequency count. Then a value is randomly generated within the selected sub-range. The number of times sub-range j has been chosen to generate a value for w_i is accumulated in the frequency counter *freq*(i, j). The main goal of the diversification generator is to create solutions that are diverse with respect to those solutions that have been already generated in the past. That is, the frequency counts work as "seed solutions" from which the diversification attempts to move away.

1. Normalize input and output data.
2. Start with $P = \emptyset$. Use the diversification method to construct a solution w between *wlow* and *whigh*. If $w \notin P$ then add w to P (i.e., $P = P \cup w$), otherwise, discard w. Repeat this step until $|P| = PSize$. Apply the improvement method to the best $b/2$ solutions in P to generate $w^{(1)}, \ldots, w^{(b/2)}$. Generate $b/2$ more solutions, where $w^{(b/2+i)} = w^{(i)}(1 + U[-0.3,0.3])$ for $i = 1, \ldots, b/2$. Build $RefSet = \{ w^{(1)}, \ldots, w^{(b)} \}$.
3. Order *RefSet* according to their objective function value such that $w^{(1)}$ is the best solution and $w^{(b)}$ the worst.

while (*NumEval* < *TotalEval*) **do**

  3. Generate *NewPairs*, which consists of all pairs of solutions in *RefSet* that include at least one new solution. Make $NewSolutions = \emptyset$.

  for (all *NewPairs*) **do**

    4. Select the next pair ($w^{(i)}, w^{(j)}$) in *NewPairs*.
    5. Obtain new solutions w as linear combinations of ($w^{(i)}, w^{(j)}$) and add them to *NewSolutions*.

  end for

  6. Select the best b solutions in *NewSolutions* and apply the improvement method.

  for (each improved w) **do**

    if (w is not in *RefSet* and $g(w) < g(w^{(b)})$) **then** make $w^{(b)} = w$ and reorder *RefSet*.

  end for

  while (*IntCount* < *IntLimit*)

    7. Make $IntCount = IntCount + 1$ and $w = w^{(1)}$.
    8. Make $w = w(1 + U[-0.05,0.05])$ and apply improvement method.

    if ($g(w) < g(w^{(1)})$) **then** Make $w^{(1)} = w$ and $IntCount = 0$.

  end while

  9. Apply the improvement method to $w^{(i)}$ for $i = 1, \ldots, b/2$ in *RefSet*.
  10. Make $w^{(b/2+i)} = w^{(i)}(1+U[-0.01,0.01])$ for $i = 1, \ldots, b/2$ in *RefSet*.

end while

Figure 7-3. Outline of training procedure.

The reference set *RefSet* is filled with the best $b/2$ solutions in P to which an improvement method is applied (see below). The *RefSet* is completed with $b/2$ more solutions generated as perturbations of the first $b/2$. The perturbation consists of multiplying each weight by $1 + U[a,b]$, where $U[a,b]$ is the uniform distribution with parameters a and b. In step 2, the solutions in *RefSet* are ordered according to quality, where the best solution is the first one in the list. In step 3, the *NewPairs* set is constructed. *NewPairs* consists of all the new pairs of solutions that can be obtained from *RefSet*, where a "new pair" contains at least one new solution. Since all the solutions are new in the initial *RefSet*, the initial *NewPairs* consists of $(b^2-b)/2$ pairs. The pairs in *NewPairs* are selected one at a time in lexicographical order to create linear combinations in step 5. We consider the following three types of linear combinations, where we assume that the reference solutions are $w^{(i)}$ and $w^{(j)}$, and r is a random number in the range (0, 1):

C1: $w = w^{(i)} - d$

C2: $w = w^{(i)} + d$ $\qquad d = r\dfrac{w^{(i)} - w^{(j)}}{2}$

C3: $w = w^{(j)} - d$

Note that we consider one convex combination (C1) and two non-convex combinations (C2 and C3) in both sides of the line joining $w^{(i)}$ and $w^{(j)}$. The following rules are used to generate solutions with these three types of linear combinations. The method generates more solutions when combines high quality solutions (intensification) that when it combines lower quality solutions (diversification):

- If $i \leq b/2$ and $j \leq b/2$ then 4 solutions are generated by applying C1 and C3 once and C2 twice.
- If $i \leq b/2$ and $j > b/2$ then 3 solutions are generated by applying C1, C2 and C3 once.
- If $i > b/2$ and $j > b/2$ then 2 solutions are generated by applying C2 once and randomly choosing between applying C1 or C3.

The solutions created as linear combinations of solutions in the reference set are added to the *NewSolutions* set. Once all combinations have been made, the best b solutions in *NewSolutions* are subjected to the improvement method in step 6. Each improved solution w is then tested for admission into *RefSet*. If a newly created solution improves upon the worst solution currently in *RefSet*, the new solution replaces the worst and *RefSet* is reordered.

The procedure now intensifies the search around the best-known solution. In step 7, the counter *IntCount* is increased and the best solution is copied to a temporary memory location w. The solution is perturbed and the improvement method is applied in step 8. The best solution is updated if the perturbation plus improvement generates a better solution. When the best solution is improved, the intensification count *IntCount* is reset. If *IntLimit* intensification iterations are performed without improving the best solution, the procedure abandons the intensification phase.

Finally, steps 9 and 10 operate on the entire *RefSet*. Step 9 applies the improvement method to the best $b/2$ solutions in *RefSet* and step 9 replaces the worst $b/2$ solutions with perturbations of the best $b/2$. The training procedure stops when the number of objective function evaluations (*NumEval*) reaches the total allowed (*TotalEval*). Note that the evaluation of the objective function $g(w)$ consists of the calculation of the mean squared error.

The SS procedure employs the well known Nelder and Mead (1965) optimizer as an improvement method. Given a set of weights w, the Nelder and Mead method starts by perturbing each weight to create an initial simplex from which to begin the local search. We use the implementation of the Nelder-Mead method in Press, et al. (1992) with the default parameters for 500 iterations. Note that this improvement method is used in three different situations during the search: (1) to improve upon the best *b*/2 solution in the initial *RefSet*, (2) to improve upon the b best solution that result from the linear combinations, (3) to improve upon the perturbed solutions generated during the intensification, and (3) to improve upon the *b*/2 best solutions when rebuilding *RefSet* in steps 9 and 10.

4. COMPUTATIONAL EXPERIMENTS

For our computational testing, we implemented, in C, the classical Back-Propagation method (BP), the extended tabu search method, ETS, of Sexton et al. (1998) and the SS method described in the previous section (SS). We run these three methods to approximate the 15 test functions introduced in El-Fallahi and Martí (2003). This test set includes well known multimodal functions such as the Six-Hump-Camelback, the Goldstein or the Schwefel instances. After preliminary experimentation we set the number of hidden neurons equal to 9 (and we keep this parameter fixed for all the functions).

The main goal in the design of an ANN is to obtain a model which makes good predictions for new inputs (i.e. to provide a good generalization). The standard way to measure how well this goal is accomplished consists of introducing an additional set of points in the domain of f called the validation set V. We assume that no point in V belongs to E (the training set) and $f(x)$ is known for all $x \in V$. Once the optimization has been performed and the weights w^* that minimize E have been set, the validation set V is presented to the resulting network and the corresponding errors are computed ($error(T,w^*)$). The network must exhibit a good fit between the target f-values and the output (prediction) in the training set and also in the validation set. In our experiments, the training set consists of 200 observations with data randomly drawn from [-100, 100] for x_1 and [-10,10] for x_2. The validation set consists of 100 additional observations drawn from the same uniform distributions.

Tables 7-1 and 7-2 report, respectively, the training and validation errors obtained with the three methods when applied to the 15 problems considered. In order to obtain statistically valid results, we ran 20 times each method on each function and report the average and standard deviation of the 20 runs (each run is limited to ten minutes). In all the cases, we have considered the same training and validation sets. All the experiments have been performed on a Pentium IV 2.8 Ghz personal computer.

Table 7-1. Training errors for three different training methods

Problem	BP	ETS	SS
1	1.60 ± 0.26	0.04 ± 0.02	0.00 ± 0.00
2	8.32 ± 4.30	1.79 ± 0.78	0.05 ± 0.03
3	1.63 ± 0.21	0.34 ±0.03	0.4 ± 0.19
4	45.52 ± 7.82	17.66 ±6	0.2 ± 0.08
5	12.62 ± 3.87	18.98 ±5.26	0.27 ± 0.26
6	13.98 ± 1.58	53.28 ±3.94	1.98 ± 0.29
7	16.09 ± 5.80	63.26 ±1.18	0.55 ± 0.01
8	0.20 ± 0.06	0.01 ±0.00	0.07 ± 0.03
9	7.35E+09±1.07E+09	3.30E+09±8.44E+07	4.95E+04±5.01E+03
10	21.40 ± 1.49	22.22± 4.12	4.6 ± 0.22
11	5.28E+06±1.34E+06	4.17E+06±1.28E+05	1.03E+03±1.07E+02
12	107.95 ± 3.01	156.12±5.57	0.1 ± 0.05
13	3.93 ± 1.97	10.13 ± 3.25	0.02 ± 0.1
14	5.58E+0 ± 6.76E+03	4.44E+04±2.48E+03	3.95E+04±2.01E+03
15	2.88 ± 0.5	527.14±3.07	0.39± 0.02

Table 7-2. Validation errors for three different training methods

Problem	BP	ETS	SS
1	1.77 ± 1.6	0.05 ± 0.02	0.00 ± 0.14
2	8.59 ± 3.94	2.13 ± 0.95	0.00 ± 0.12
3	1.68 ± 0.22	0.48 ± 0.05	0.44 ± 0.43
4	43.89 ± 9.86	21.71 ± 7.49	8.76 ± 26.13
5	14.30 ± 5.50	19.85 ± 6.19	1.06 ± 2.42
6	15.32 ± 0.87	51.54 ± 6.07	106.11 ± 221.11
7	21.56 ± 11.97	60.62 ± 5.31	52.04 ± 137.21
8	0.19 ± 0.06	0.04 ± 0.0	0.01 ± 0.02
9	1.22E+10±1.24E+04	7.03E+09±5.38E+08	3.3049E+10±5.04E+10
10	13.67 ± 0.02	16.59 ± 0.57	107.24 ± 192.39
11	4.13E+06±4.02E+03	6.51E+06±6.02E+05	5.70E+07±1.33E+08
12	111.17 ± 6.93	149.31 ± 7.22	0.03 ± 0.04
13	5.25 ± 2.56	10.32 ± 0.60	0.10 ± 0.70
14	5.64E+04±2.62E+03	4.33E+04±1.94E+03	1.30E+07± 1.30E+07
15	2.93 ± 0.53	554.68 ± 22.76	275.41±173.12

This experiment shows that none of the methods can effectively handle problems 9, 11 and 14 within the run time considered. (We also ran the training methods in these three instances for about half an hour of CPU time with no significant improvement.) Therefore, we can say that in practice the neural network configuration that we tested is not able to approximate these functions.

Considering the average values over the 20 runs, Table 7-1 shows that the SS method is able to obtain the best solutions with respect to the training error in 13 instances, while the ETS method obtains 2 best solutions. Table 7-2 shows similar results, with SS obtaining 10 best solutions with respect to the testing error and ETS obtaining 5 (including instances 9, 11 and 14). It must be noted that the results of the BP method are good considering ots simplicity since it is able to obtain the best solutions, with respect to the validation error, in 4 cases (instances 6, 7 10 and 15). In those examples the other methods present significantly larger error values. Therefore, we can conclude that a hybrid method of both, metaheuristics and back-propagation based algorithms, could be the core of an advanced training method for future developments.

5. CONCLUSIONS

In this chapter we have described the implementation of scatter search for training a single-layer feed-forward neural network. The focus of scatter search leads to a number of implications associated with the design of improved optimization procedures. These research opportunities carry with them an emphasis on producing systematic and strategically designed rules, rather than following the policy of relegating decisions to random choices, as often is fashionable in evolutionary methods. Our goal was to develop a training procedure to obtain neural networks capable of providing high-quality predictions. Our experiments show that the scatter search implementation reaches a prediction accuracy that in most cases is acceptable. The experimentation also shows that the SS implementation improves upon the results found with a recently developed tabu search method.

ACKNOWLEDGMENTS

Research by Rafael Martí is partially supported by the *Ministerio de Educación y Ciencia* (refs. TIN2004-20061-E and TIC2003-C05-01) and by the *Agencia Valenciana de Ciència i Tecnologia* (ref. GRUPOS03 /189).

REFERENCES

El-Fallahi, A., and Martí, R., 2003, Tabu and Scatter Search for Training Neural Networks, in: *Computational Modeling and Problem Solving in the Networked World*, H. K. Bhargava, and N. Ye, ed., Interfaces in Computer Science and Operations Research, pp. 79-96.

Glover F., 1963, Parametric combinations of local job shop rules, *ONR Research Memorandum*. **117**, GSIA, Carnegie Mellon University, Pittsburgh, PA.

Glover, F., 1977, Heuristics for integer programming using surrogate constraints, *Decision Sciences* **8**: 156-166.

Glover, F., 1994, Tabu search for nonlinear and parametric optimization (with links to genetic algorithms), *Discrete Applied Mathematics* **49**: 231-255.

Glover, F., 1998, A template for scatter search and path relinking, in: *Artificial Evolution, Lecture Notes in Computer Science* 1363, J. K. Hao, E. Lutton, E. Ronald, M. Schoenauer, D. Snyers, eds., Springer, pp. 13-54.

Glover, F., 2004, Parametric tabu search for mixed integer programming, Technical Report, University of Colorado at Boulder; http://spot.colorado.edu/ ~glover/Recentpapers.html.

Glover, F., and Laguna, M., 1997, *Tabu Search*, Kluwer Academic Publishers, Boston.

Glover, F., Laguna, M., and Martí, R., 2000, Fundamentals of scatter search and path relinking, *Control and Cybernetics* **29**(3): 653-684.

Laguna, M., and Martí, R., 2000, Neural network prediction in a system for optimizing simulations, *IIE Transaction* **34**(3): 273-282.

Laguna, M., and Martí, R., 2003, *Scatter Search – Methodology and Implementations in C*, Kluwer Academic Publishers, Boston.

Nelder, J. A., and Mead, R., 1965, A simplex method for function minimization, *Computer Journal* **7**: 308-313.

Press, W. H., Teukolsky, S. A., Vetterling, W. T., and Flannery, B. P., 1992, *Numerical Recipes: The Art of Scientific Computing*, Cambridge University Press (www.nr.com).

Sexton, R. S., Alidaee, B., Dorsey, R. E., and Johnson, J. D., 1998, Global optimization for artificial neural networks: A tabu search application, *European Journal of Operational Research* **106**: 570-584.

Chapter 8

ANT COLONY OPTIMIZATION

Krzysztof Socha[1] and Christian Blum[2]
[1]*IRIDIA, Université Libre de Bruxelles, Brussels, Belgium;*
[2]*ALBCOM, LSI, Universitat Politècnica de Catalunya, Barcelona Spain*

Abstract: Ant colony optimization (ACO) is a metaheuristic that was originally introduced for solving combinatorial optimization problems. In this chapter we present the general description of ACO, as well as its adaptation for the application to continuous optimization problems. We apply this adaptation of ACO to optimize the weights of feed-forward neural networks for the purpose of pattern classification. As test problems we choose three data sets from the well-known PROBEN1 medical database. The experimental results show that our algorithm is comparable to specialized algorithms for feed-forward neural network training. Furthermore, the results compare favourably to the results of other general-purpose methods such as genetic algorithms.

Key words: Ant colony optimization, continuous optimization, pattern classification, feed-forward neural network training.

1. INTRODUCTION

In the early 90's, ant colony optimization (ACO) (Dorigo, 1992; Dorigo *et al.*, 1991; Dorigo *et al.*, 1996) was introduced as a novel nature-inspired metaheuristic for solving hard combinatorial optimization (CO) problems. According to Papadimitriou and Steiglitz (1982), a CO problem $P=(S,f)$ is an optimization problem in which there is given a finite set of solutions S (also called *search space*) and an objective function $f: S \rightarrow R^{+}$ [2] that assigns a positive cost value to each of the solutions. The goal is either to find a

[2] R^{+} denotes the space of nonnegative real values.

solution of minimum cost value[3], or—as in the case of approximate solution techniques—a good enough solution in a reasonable amount of time. ACO algorithms follow—as do all metaheuristics—the latter goal.

Examples of metaheuristics other than ACO are tabu search (Glover, 1989; Glover, 1990; Glover and Laguna, 1997), simulated annealing (Kirkpatrick *et al.*, 1983; Černý, 1985), and evolutionary computation (Fogel *et al.*, 1966; Rechenberg, 1973; Holland, 1975; Goldberg, 1989). For more general literature on metaheuristics, see (Glover and Kochenberger, 2002; Blum and Roli, 2003; Hoos and Stützle, 2004). Some metaheuristics have been designed with continuous optimization in mind, e.g. evolutionary strategies (ES). However, most metaheuristics were originally introduced to solve CO problems, and only recently their adaptation to solve continuous optimization problems enjoys increasing attention. In continuous optimization we generally want to find a vector $x^* \in K$ (where K is an n-dimensional subspace of R^n) such that $f(x^*) \leq f(x)$, for all $x \in K$ and $f : K \rightarrow R^+$. Examples of such adapted metaheuristics are simulated annealing algorithms (Siarry *et al.*, 1997), or differential evolution (Storn and Price, 1997; Chelouah and Siarry, 2000; Chelouah and Siarry, 2003) from the evolutionary computation field. Among the CO oriented metaheuristics, tabu search algorithms as, for example, (Battiti and Tecchioli, 1996) were among the first to be applied to continuous problems. Some of the above cited methods are hybrids that make use of the well-known Nelder-Mead simplex algorithm for continuous optimization (Nelder and Mead, 1965).

Among the most recent applications of metaheuristics to continuous problems are ant-related algorithms (Mathur *et al.*, 2000; Dréo and Siarry, 2002; Socha, 2004). Only the most recent one of these attempts – namely (Socha, 2004) – can be labelled ant colony optimization. The other attempts rather loosely follow the idea of ant-based algorithms. In this chapter we first present the general idea of ACO. Then, we describe the attempts of creating ant algorithms for continuous optimization. Finally, we describe our approach, which is an extended version of the algorithm proposed by Socha (2004). We apply this algorithm to the training of feed-forward neural networks (NNs). To test our algorithm we apply it to three benchmark classification problems from the medical field: the diagnosis of breast cancer, the diagnosis of diabetes, and the diagnosis of heart disease.

[3] Note that minimizing over an objective function f is the same as maximizing over $-f$. Therefore, every CO problem can be described as a minimization problem.

The remaining part of this chapter is organized as follows. Section 2 presents the ant colony optimization metaheuristic. In particular, subsection 2.1 gives an overview of the basics of ACO for combinatorial optimization, and subsection 2.2 summarizes the existing approaches of applying ant algorithms to continuous optimization problems. Section 3 describes our ACO algorithm for continuous optimization, henceforth denoted by $ACO_{\mathbb{R}}$. Following this, Section 4 presents in detail the specific problem of training feed-forward NNs, the experimental setup, and the results obtained. Finally, Section 5 concludes the chapter.

2. ANT COLONY OPTIMIZATION: THE GENERAL IDEA

ACO algorithms have their origins in a field known as swarm intelligence (SI) (Bonabeau *et al.*, 1999). SI algorithms take their inspiration from the collective behaviour of, for example, social insects, flocks of birds, or fish schools. Examples include algorithms for clustering and data mining inspired by ants' cemetery building behaviour, dynamic task allocation algorithms inspired by the behaviour of wasp colonies, particle swarm optimization (PSO) algorithms, and many more. The inspiring source of ACO is the foraging behaviour of real ants. When searching for food, ants initially explore the area surrounding their nest in a random manner. As soon as an ant finds a food source, it evaluates the quantity and the quality of the food and carries some of it back to the nest. During the return trip, the ant deposits a chemical pheromone trail on the ground. The quantity of pheromone deposited, which may depend on the quantity and quality of the food, will guide other ants to the food source. As it has been shown in (Deneubourg *et al.*, 1990), indirect communication between the ants via pheromone trails enables them to find shortest paths between their nest and food sources. This characteristic of real ant colonies is exploited in artificial ant colonies in order to solve optimization problems.

```
while termination conditions not met do
    ScheduleActivities
        AntBasedSolutionConstruction()
        PheromoneUpdate()
        DaemonActions()          {optional}
    end ScheduleActivities
end while
```

Figure 8-1. Framework of the ant colony optimization (ACO) metaheuristic.

2.1 Ant Colony Optimization for CO Problems

The central component of an ACO algorithm is a parameterized probabilistic model, which is called the *pheromone model*. The pheromone model consists of a vector of model parameters T called *pheromone trail parameters*. The pheromone trail parameters $T_i \in T$, which are usually associated with components of solutions, have values τ_i, called *pheromone values*. The pheromone model is used to probabilistically generate solutions to the problem under consideration by assembling them from a finite set of solution components. At run-time, ACO algorithms update the pheromone values using previously generated solutions. The update aims to concentrate the search in regions of the search space containing high quality solutions. In particular, the reinforcement of solution components depending on the solution quality is an important ingredient of ACO algorithms. It implicitly assumes that good solutions consist of good solution components. To learn which components contribute to good solutions can help assembling them into better solutions. In general, the ACO approach attempts to solve an optimization problem by repeating the following two steps:

- candidate solutions are constructed using a pheromone model, that is, a parameterized probability distribution over the solution space;
- the candidate solutions are used to modify the pheromone values in a way that is deemed to bias future sampling toward high quality solutions.

The ACO metaheuristic framework is shown in Figure 8-1. It consists of three algorithmic components that are gathered in the ScheduleActivities construct. The ScheduleActivities construct does not specify how these three activities are scheduled and synchronized. This is up to the algorithm designer. In the following we explain these three algorithm components in more detail.

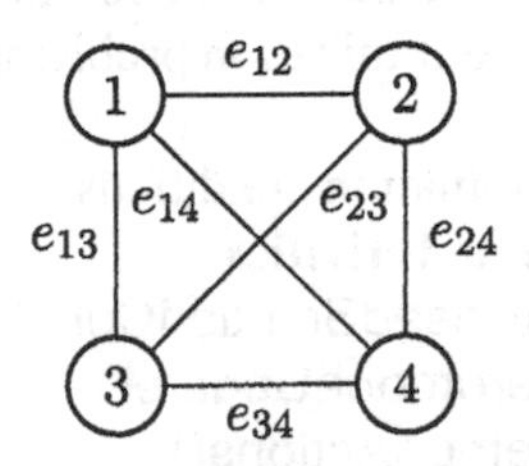

Figure 8-2. A small TSP problem instance on 4 cities in form of a graph $G=(V,E)$. The edges e_{ij} that connect the cities have associated distances d_{ij}.

AntBasedSolutionConstruction(): Artificial ants are constructive heuristics that assemble solutions as sequences of solution components taken from a finite set of solution components $C=\{c_1,...,c_n\}$. Solution construction starts with an empty partial solution s^p=<>. Then, at each construction step the current partial solution s^p is extended by adding a feasible solution component from the set $N(s^p) \in C \setminus s^p$, which is defined by the solution construction mechanism. The process of constructing solutions can be regarded as a walk (or a path) on the so-called *construction graph* (see Figure 8-3 for an example) $G_C=(C,L)$ whose vertices are the solution components C and the set L are the connections. The allowed walks on G_C are hereby implicitly defined by the solution construction mechanism that defines set $N(s^p)$ with respect to a partial solution s^p. The choice of a solution component from $N(s^p)$ is at each construction step done probabilistically with respect to the pheromone model T, which consists of *pheromone trail parameters* $T_i \in T$ that are associated to components $c_i \in C$.[4] The values of these parameters—the *pheromone values*—are denoted by τ_i. In most ACO algorithms the probabilities for choosing the next solution component—also called the *transition probabilities*—are defined as follows:

$$p(c_i \mid s^p) = \frac{\tau_i^{\alpha} \cdot \eta(c_i)^{\beta}}{\sum_{c_j \in N(s^p)} \tau_j^{\alpha} \cdot \eta(c_j)^{\beta}}, \forall c_i \in N(s^p) , \quad (1)$$

where η is a weighting function, which is a function that, sometimes depending on the current partial solution, assigns at each construction step a heuristic value $\eta(c_i)$ to each feasible solution component $c_i \in N(s^p)$. The values that are given by the weighting function are commonly called the *heuristic information*. Furthermore, α and β are positive parameters whose values determine the relation between pheromone information and heuristic information. In Figure 8-3 is given an example of the solution construction for the small travelling salesman (TSP) problem instance that is shown is given in Figure 8-2.

[4] Note that the description of the ACO metaheuristic as given for example in (Dorigo and Stützle, 2004) allows also connections of the construction graph to be associated with a pheromone trail parameter. However, for the purpose of this introduction it is sufficient to assume that pheromone trail parameters are associated with components.

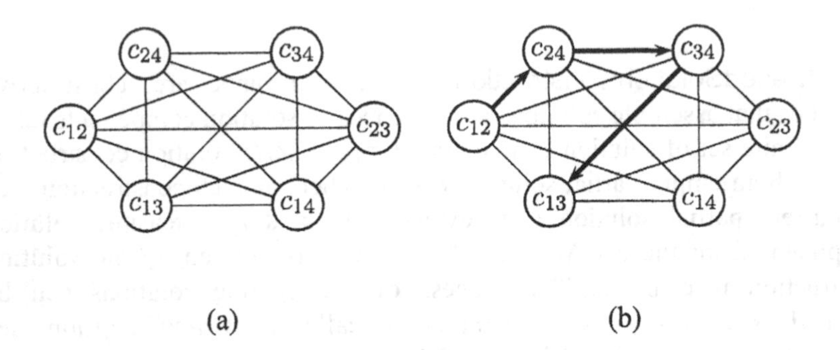

Figure 8-3. (a) shows the construction graph $G_C=(C,L)$ with respect to the TSP instance shown in Figure 8-2. The set of solution components C consists of a solution component c_{ij} for each edge e_{ij}, and the pheromone model consists of a pheromone trail parameter T_{ij} for each solution component c_{ij}. As heuristic information we choose the inverse of the distances between the cities. Therefore, it holds that $\eta(c_{ij})=1/d_{ij}$ for all c_{ij}. The construction mechanism is as follows. In the first construction step, any solution component can be chosen. For the remaining construction steps the set of solution components is restricted so that the sequence of solution components always corresponds to a path in G (respectively, to a Hamiltonian cycle in G after the last construction step). (b) shows a path on the construction graph that corresponds to the construction of solution $s=<c_{12},c_{24},c_{34},c_{13}>$.

PheromoneUpdate(): Different ACO algorithms—such as, for example, Ant Colony System (ACS) (Dorigo and Gambardella, 1997) and *MAX–MIN* Ant System (*MM*AS) (Stützle and Hoos, 2000)—mainly differ in the update of the pheromone values they apply. In the following, we outline a common pheromone update rule in order to provide the general idea. This pheromone update consists of two parts. First, a *pheromone evaporation*, which proportionally decreases all the pheromone values, is performed. From a practical point of view, pheromone evaporation is needed to avoid an overly rapid convergence of the algorithm toward a sub-optimal region. It implements a useful form of *forgetting*, favouring the exploration of new areas in the search space. Second, one or more solutions from the current and/or from earlier iterations are used to increase the values of pheromone trail parameters on solution components that are part of these solutions:

$$\tau_i \leftarrow (1-\rho)\tau_i + \rho \sum_{\{s \in S_{upd} | c_i \in s\}} F(s) \quad , \tag{2}$$

for $i=1,\dots,n$. Here, S_{upd} is the set of solutions that are used for the update. Furthermore, $\rho \in (0,1]$ is a parameter called evaporation rate, and $F : S \rightarrow R^+$ is a function such that $f(s) < f(s') \Rightarrow F(s) \geq F(s')$, $\forall\, s \neq s' \in S$. $F(\cdot)$ is commonly called the *quality function*. Instantiations of this update rule are obtained by different specifications of S_{upd}, which—in many cases—is a subset of $S_{iter} \cup \{s_{bs}\}$, where S_{iter} is the set of solutions that were constructed in the current iteration, and s_{bs} is the best-so-far solution, that is, the best solution found since the first algorithm iteration. A well-known example is the *AS-update* rule, that is, the update rule of Ant System (AS) (Dorigo *et al.*, 1996). The AS-update rule is obtained from update rule 2 by setting

$$S_{upd} \leftarrow S_{iter} \quad .$$

This update rule is well-known due to the fact that AS was the first ACO algorithm to be proposed in the literature. An example of a pheromone update rule that is more used in practice is the *IB-update* rule (where IB stands for *iteration-best*). The IB-update rule is given by:

$$S_{upd} \leftarrow \arg\max\{F(s) \mid s \in S_{iter}\}$$

The IB-update rule introduces a much stronger bias towards the good solutions found than the AS-update rule. However, this increases the danger of premature convergence. An even stronger bias is introduced by the *BS-update* rule, where BS refers to the use of the best-so-far solution s_{bs}. In this case, S_{upd} is set to $\{s_{bs}\}$. In practice, ACO algorithms that use variations of the IB-update or the BS-update rule include mechanisms to avoid premature convergence and tend to achieve better results than algorithms that use the AS-update rule. Examples are ACS and *MM*AS as mentioned above.

DaemonActions(): Daemon actions can be used to implement centralized actions which cannot be performed by single ants. Examples are the application of local search methods to the constructed solutions, or the collection of global information that can be used to decide whether it is useful or not to deposit additional pheromone to bias the search process from a non-local perspective. As a practical example, the daemon may decide to deposit extra pheromone on the solution components that belong to the best solution found so far.

After the initial proof-of-concept application to the travelling salesman problem (TSP) (Dorigo *et al.*, 1991; Dorigo *et al.*, 1996), ACO was applied to many other CO problems. Examples include the applications to assignment problems (Costa and Hertz, 1997; Maniezzo and Colorni, 1999; Maniezzo, 1999; Socha et al., 2003; Stützle and Hoos, 2000), scheduling problems (Stützle, 1998; den Besten *et al.*, 2000; Gagné *et al.*, 2002; Merkle *et al.*, 2002; Blum and Sampels, 2004), and vehicle routing problems (Gambardella *et al.*, 1999; Reimann *et al.*, 2004). Among other applications, ACO algorithms are currently state-of-the-art for solving the sequential ordering problem (SOP) (Gambardella and Dorigo, 2000), the resource constraint project scheduling (RCPS) problem (Merkle *et al.*, 2002), and the open shop scheduling (OSS) problem (Blum, 2005). For an overview of applications of ACO we refer the interested reader to (Dorigo and Stützle, 2004).

2.2 Ant-Related Algorithms for Continuous Optimization

As indicated in the previous section, ACO has proven to be an efficient and versatile tool for solving various combinatorial optimization problems. In recent years some attempts were also made to use them for tackling continuous optimization problems. However, a direct application of the ACO metaheuristic to continuous domains is not straightforward. Hence, the early proposals often drew inspiration from ACO, but they did not follow exactly the same methodology. This has changed since the publication of the work by Socha (2004), which is the first real ACO algorithm for continuous optimization. In the following paragraphs we highlight the main characteristics of the various ant algorithms proposed for continuous domains.

One of the first attempts to apply ant-based ideas to continuous optimization problems was Continuous ACO (CACO) (Bilchev and Parmée, 1995). In CACO the ants start from a *nest* situated somewhere in the search space. The artificial pheromone information is kept as numerical values that are each assigned to a vector. At each iteration of the algorithm, the ants choose probabilistically the vector from which they then continue the search. This is followed by some random moves. The vectors are updated with the best results found. Although the authors of CACO state to have taken the inspiration for their algorithm from the original ACO formulation, there are some important differences. There is a new notion of the *nest* introduced, which does not exist in the ACO metaheuristic. Also, CACO does not perform an incremental construction of solutions, which is one of the main characteristics of the ACO metaheuristic.

Another ant-related approach to continuous optimization is the API algorithm (Monmarché *et al.*, 2000). The API algorithm does not use any artificial pheromone information, neither do the authors claim that API is based on the ACO metaheuristic. In API, the ants perform their search independently, but starting from the same nest, which is periodically moved to other locations. The ants use a mechanism labelled *tandem running* as a type of recruitment strategy. API is the only ant algorithm published so far–apart from the later work of Socha (2004)—that allows tackling both discrete and continuous optimization problems.

The third ant-based approach to continuous optimization is Continuous Interacting Ant Colony (CIAC) (Dréo and Siarry, 2002). In contrast to a standard ACO algorithm, CIAC uses two types of communication between ants: artificial pheromone information (i.e., spots of pheromone deposited in the search space) and direct communication between ants. Furthermore, CIAC is not performing an incremental construction of solutions, which is a vital part of standard ACO algorithms.

Finally, Socha (2004) proposed ACO*, an ACO algorithm for continuous domains that follows closely the spirit of ACO algorithms for CO problems. In particular, ACO* is also based on step-by-step construction of solutions. The main idea of this algorithm is as follows. In ACO algorithms for CO problems, each solution construction step concerns the (probabilistic) choice of a solution component from a set of allowed solution components (see Section 2.2). Hereby, each solution component corresponds to the assignment of a certain value to one of the decision variables. The choice of a solution component is at each step performed probabilistically according to Equation 1, which defines a discrete probability distribution. In contrast, in ACO* the construction of a solution works by choosing for each of the continuous variables a domain value by sampling a so-called probability density function. Therefore, the main idea of ACO* is to replace discrete distributions with continuous distributions (see Figure 8-4). In that sense, ACO* is closely related to so-called estimation of distribution algorithms (EDAs). See, for example, (Kern et al., 2004).

3. $ACO_{\mathbb{R}}$ FOR CONTINUOUS OPTIMIZATION

In this section, we present the $ACO_{\mathbb{R}}$ algorithm—an extension of the idea of ACO* (Socha, 2004). In the following we assume to tackle a continuous optimization problem of the following form. Given are n decision variables $X=\{X_1,...,X_n\}$ with continuous domains $\{D_1,...,D_n\}$. For constructing a solution, the algorithm chooses for each of the n decision variables a domain value by sampling probability density functions (PDFs).

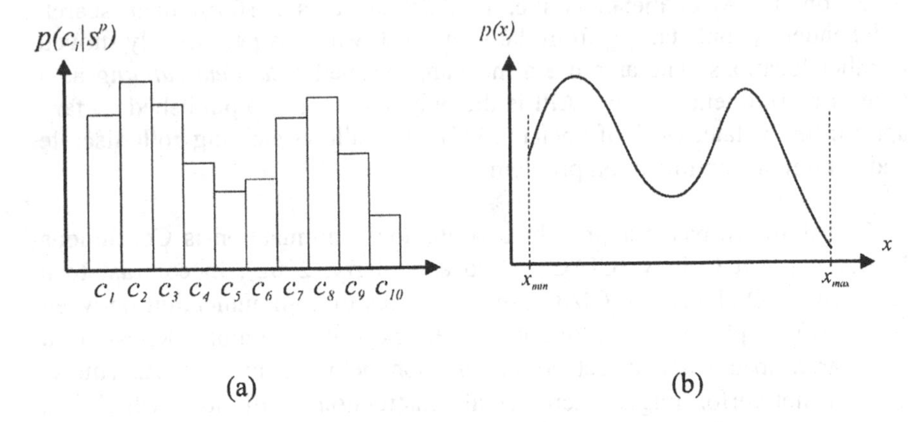

Figure 8-4. (a) The discrete probability distribution that is employed at each construction step when combinatorial problems are concerned, and (b) a continuous probability density function for continuous problems. Hereby, x_{min} and x_{max} respresent the minimal, respectively the maximal allowed domain values.

These PDFs are, for each solution construction (and for each decision variable), produced from a population P of solutions that the algorithm keeps at all times. The management of this population works as follows. Before the start of the algorithm, the population—whose size k is a parameter of the algorithm—is filled with randomly generated solutions. At each iteration a set of m solutions is generated by m ants and added to P. Then, the solutions of this extended population are ranked according to their objective function values, and the m worst solutions are removed. This mechanism biases the search process towards the best solutions found during the search process, and keeps the population size fixed to m at all times. Note that the population P of solutions takes over the role of the pheromone information that is used in ACO algorithms for CO problems as a storage of search experience. A similar approach has been used before by Guntsch and Middendorf (2002) in case of Population-Based ACO for CO problems.

For constructing a solution, an ant acts as follows. First, it transforms the original set of decision variables $\boldsymbol{X}$ into a set of temporary variables $\boldsymbol{Z}=\{Z_1, \ldots, Z_n\}$. The purpose of introducing temporary variables is to improve the algorithms performance by limiting the correlation between decision variables. Note that this transformation also affects the population P of solutions: All the solutions are transformed to the new coordinate system as well. The method of transforming the set of decision variables is presented towards the end of this section.

At each of the n construction steps $i=1,...,n$, the ant chooses a value for the corresponding decision variable Z_i. For performing this choice it uses a so-called Gaussian kernel PDF, which is a weighted superposition of several Gaussian functions. For a decision variable Z_i the corresponding Gaussian kernel G_i is given as follows:

$$G_i(z) = \sum_{j=1}^{k} \omega_j \cdot g_j(z) = \sum_{j=1}^{k} \omega_j \cdot \frac{1}{\sigma_j \sqrt{2\pi}} e^{-\frac{(z-\mu_j)^2}{2\sigma_j^2}} \text{, for all } z \in R \quad (3)$$

where the j-th Gaussian function g_j is derived from the j-th member of population P. Remember that k is the number of Gaussian functions composing the Gaussian kernel PDF. Note that $\boldsymbol{\omega}$, $\boldsymbol{\mu}$, and $\boldsymbol{\sigma}$ are vectors of size k. $\boldsymbol{\omega}$ is the vector of weights, whereas $\boldsymbol{\mu}$ and $\boldsymbol{\sigma}$ are the vectors of means and standard deviations respectively. Figure 8-5 presents an example of a Gaussian kernel PDF consisting of five separate Gaussian functions.

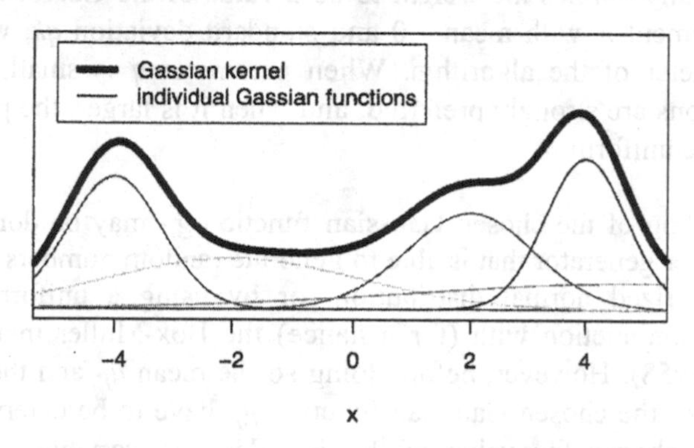

Figure 8-5. An example of a Gaussian kernel PDF consisting of five separate Gaussian functions.

Sampling directly the Gaussian kernel PDF as defined in Equation (3) is problematic. It can, however, be accomplished by the following procedure, which can be proven to be equivalent to sampling G_i directly.

Before starting the construction of a solution, each ant chooses exactly one of the k Gaussian functions, and uses this Gaussian function, henceforth denoted by g_{j^*}, for all n construction steps. The Gaussian function g_{j^*} is chosen with following probability distribution:

$$p_j = \frac{\omega_j}{\sum_{l=1}^{k} \omega_l}, \tag{4}$$

where ω_j is the weight of the j-th Gaussian function g_j, which is obtained as follows. All solutions in P are ranked with respect to their quality; with the best solution having rank 1. Assuming the rank of the j-th solution in P to be r, the weight ω_j of the j-th Gaussian function is calculated according to the following formula:

$$\omega_j = \frac{1}{qk\sqrt{2\pi}} e^{-\frac{(r-1)^2}{2q^2k^2}}, \tag{5}$$

which essentially defines the weight to be a value of the Gaussian function with the argument r, with mean 1.0 and standard deviation qk, where q is also a parameter of the algorithm. When parameter q is small, the best-ranked solutions are strongly preferred, and when it is larger, the probability becomes more uniform.

The sampling of the chosen Gaussian function g_{j^*} may be done using a random number generator that is able to generate random numbers according to a parameterized normal distribution, or by using a uniform random generator in conjunction with (for instance) the Box-Muller method (Box and Muller, 1958). However, before doing so the mean μ_{j^*} and the standard deviation σ_{j^*} of the chosen Gaussian function g_{j^*} have to be determined. As mean μ_{j^*} we choose the value of the i-th decision variable in the j^*-th solution. It remains to specify the standard deviation σ_{j^*}. In order to establish the value of this standard deviation we calculate the average distance of the other population members from the j^*-th solution (in dimension i) and multiply it by the parameter ρ, which determines the speed of convergence:

$$\sigma_{j^*} = \frac{1}{k-1} \rho \sum_{l=1}^{k} | z_i^l - z_i^{j^*} | \tag{6}$$

Parameter ρ has a role similar to the pheromone evaporation rate ρ in ACO for CO problems. The higher the value of $\rho>0$, the lower the convergence speed of the algorithm, and hence the lower the learning rate. Since this whole process is done for each dimension (i.e., each decision variable) in turn, each time the distance is calculated only with the use of one single dimension (the rest of them are discarded). This ensures that the algorithm is able to adapt convergence, but also allows the handling of problems that are scaled differently in different directions.

Next, we describe how the set of temporary decision variables $\boldsymbol{Z}$ is created from the original set $\boldsymbol{X}$. Note that ACO algorithms in general do not exploit correlation information between different decision variables (or components). In $ACO_{\mathbb{R}}$, due to the specific way the pheromone is represented (i.e., as the population of solutions), it is in fact possible to take into account the correlation between the decision variables. An obvious choice for adapting the coordinate system to the distribution of population P is the Principal Component Analysis (PCA) (Hastie *et al.*, 2001). Although PCA works very well for reasonably regular distributions, its performance is no longer that interesting in case of more complex functions. The mechanism that we designed instead, is relatively simple. Each ant at each step of the construction process chooses a new direction. This direction is chosen by randomly selecting a solution s_u from P that is reasonably far away from the j^*-th solution chosen for defining the Gaussian function g_{j^*}. Then, the vector from the u-th solution to the j^*-th solution becomes the new direction. The probability of choosing the u-th solution is the following:

$$p(s_u \mid s_{j^*}) = \frac{d(s_u, s_{j^*})^4}{\sum_{l=1}^{k} d(s_l, s_{j^*})^4}, \tag{7}$$

where function $d(.,.)$ returns the distance between two members of the population P. Once this new direction is chosen, the new orthogonal basis for the ant's coordinate system is created using the Gram-Schmidt process (Golub and van Loan, 1989). It takes as input all the (already orthogonal) directions chosen in earlier construction steps and the newly chosen vector. The remaining missing vectors (for the remaining dimensions) are chosen randomly. Then, all the current coordinates of all the solutions in the population are rotated and recalculated according to this new orthogonal base resulting in the set of new temporary variables $\boldsymbol{Z}$. Only then is the ant able to measure the average distance, and subsequently to sample from the PDF (as it can now calculate the mean and standard deviation). At the end of the construction process, the chosen values of the temporary variables $\boldsymbol{Z}$ are converted back into the original coordinate system $\boldsymbol{X}$.

Finally, we deal with the subject of constraint handling. Note that the way of generating new solutions as explained above, might lead to unfeasible solutions. In the literature on evolutionary algorithms we find several ways to deal with unfeasible solutions, including rejection, repair, or penalization. All these methods can also be applied in $ACO_{\mathbb{R}}$.

4. EXPERIMENTAL SETUP AND RESULTS

After the presentation of $ACO_{\mathbb{R}}$ we will now outline in detail the application of $ACO_{\mathbb{R}}$ to the training of feed-forward NNs for the purpose of pattern classification, as well as the experimental setup and the results that we obtained.

4.1 The Problem

Due to their practical importance, we chose to evaluate the performance of $ACO_{\mathbb{R}}$ on classification problems arising in the medical field. More specifically, we chose three problems from the well-known PROBEN1[5] benchmark set (Prechelt, 1994), namely Cancer1, Diabetes1, and Heart1. Each of these problems consists of a number of patterns together with their correct classification, that is, Cancer1 consists of 699 patterns from a breast cancer database, Diabetes1 consists of 768 patterns concerning diabetes patients, and Heart1 is the biggest of the three data sets, consisting of 920 patterns describing a heart condition. Each pattern of the three problems is either classified as pathological, or as normal. Furthermore, each pattern consists of a number of measurements (i.e., numerical values): 9 measurements in the case of Cancer1, 8 in the case of Diabetes1, and 35 in the case of Heart1. The goal consists in generating a classifier that takes the measurements of a pattern as input, and provides its correct classification as output.

Feed-forward neural networks (NNs) are popular classification tools. Each feed-forward NN consists of an input layer of neurons. In case of the classification problem the input layer consists of as many neurons as there are measurements in the patterns, that is, for each measurement there exists exactly one input neuron. Furthermore, a feed-forward NN consists of an arbitrary number of hidden layers of neurons, and an output layer (for an example, see Figure 8-6). The output layer consists of as many neurons as the data set has classes. In our case, the output layer consists of 2 output neurons. Given the weights of all the neuron connections, in order to classify a pattern, one provides its measurements as input to the input neurons,

[5] It is available online at: ftp://ftp.ira.uka.de/pub/neuron/proben1.tar.gz.

propagates the output signals from layer to layer until the output signals of the output neurons are obtained. Each output neuron is identified with one of the possible classes. The output neuron that produces the highest output signal classifies the respective pattern (winner takes all).

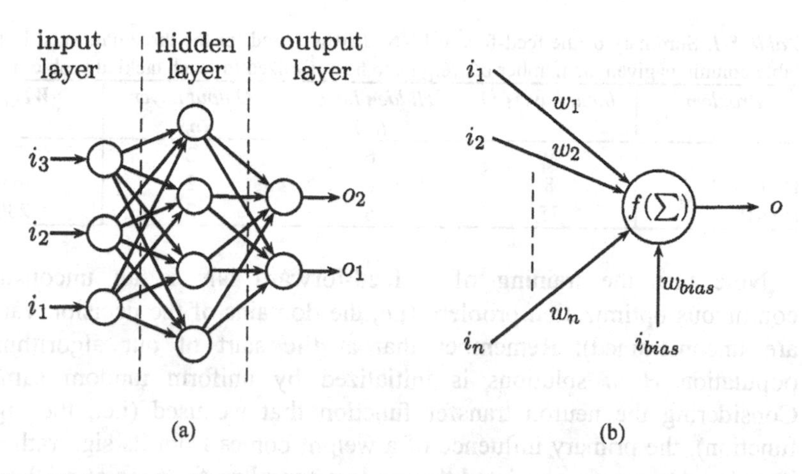

Figure 8-6. (a) shows a feed-forward NN with one hidden layer. Note that all the neurons of each layer are connected to all the neurons of the next layer. (b) shows one single neuron (from either the hidden layer, or the output layer). The neuron receives inputs (i.e., signals i_l, weighted by weights w_l) from each neuron of the previous layer. Additionally, it receives a so-called bias input i_{bias} with weight w_{bias}. The transfer function *f()* of a neuron transforms the sum of all the weighted inputs into an output signal, which servers as input for all the neurons of the following layer. Input and output signals, biases, and weights are real values.

The process of generating a NN classifier consists of determining the weights of the connections between the neurons such that the NN classifier shows a high performance. Since the weights are real-valued, this is a continuous optimization problem.

Concerning the hidden neuron layers of the feed-forward NNs that we used, we took inspiration from the literature. More specifically we used the same structure of hidden layers that were used in (Alba and Chicano, 2004). For an overview of the feed-forward NNs that we used see Table 8-1. The number of weights to be optimized is—for each of the three data sets—given by the following formula:

$$n_h(n_i+1)+n_o(n_h+1) \quad , \tag{8}$$

where n_i, n_h, and n_o are respectively the numbers of input, hidden, and output neurons. Note that the additional input for each neuron of the hidden layer and the output layer represents the *bias* inputs. The last column of Table 8-1 provides the number of weights to be optimized.

Table 8-1. Summary of the feed-forward NNs that we used for the three data sets. In the last table column is given the number of weights to be optimized for each tackled problem

Problem	*Input Layer(n_i)*	*Hidden Layer (n_h)*	*Output Layer (n_o)*	*Weights*
Cancer1	9	6	2	74
Diabetes1	8	6	2	68
Heart1	35	6	2	230

Note that the training of a feed-forward NN is an unconstrained continuous optimization problem (i.e, the domains of the decision variables are unconstrained). Remember that at the start of our algorithm, the population P of solutions is initialized by uniform random sampling. Considering the neuron transfer function that we used (i.e., the sigmoid function), the primary influence of a weight comes from its sign rather than its value. Hence, we restricted the random sampling for generating the initial solution to the interval *[-1,1]*.[6]

4.2 Training and Solution Evaluation

Neural networks for pattern classification are usually expected to exhibit a generalization capability, that is, new patterns that were not used for the training of the neural network should also be classified correctly. Having this objective in mind, the training of neural network classifiers works generally as follows. First, the set of patterns is divided into training set and test set. In our case we chose randomly 75% of all available patterns of a problem as training set (denoted by P^*), and the remaining 25% of the patterns as test set (denoted by P').

For the training of the weights of a feed-forward NN, a function is needed that distinguishes between different solutions. In other words, we need a function that measures the classification power of a solution, that is, a weight setting, with respect to the training set. For this purpose, we have used the function that is routinely used for this purpose, namely the Square Error Percentage (SEP):

[6] This restriction applies only to the initial interval used. During the search, the ACO_R algorithm can sample values outside of this initial interval.

$$SEP = 100 \frac{o_{\max} - o_{\min}}{n_o \, |P^*|} \sum_{p=1}^{|P^*|} \sum_{i=1}^{n_o} \left(t_i^p - o_i^p \right)^2 \quad , \tag{9}$$

where o_{max} and o_{min} are respectively the maximum and minimum values of the output signals of the output neurons (depeding on the neuron transfer function), n_O is the number of output neurons, and t^p_i and o^p_i represent respectively the expected and actual values of output neuron i for pattern p.

Finally, in order to assess the quality of the final solution found by a given algorithm, we used the Classification Error Percentage (CEP) as the performance measure. CEP represents the percentage of incorrectly classified patterns from the test set.

4.3 Algorithms Used for Comparison

The goal of our experimentation was to evaluate whether $ACO_{\mathbb{R}}$ may be used for training feed-forward NNs, and if so, we were interested in how it would compare to other algorithms. In order to be able to draw any meaningful conclusions, it is required to have some reference algorithm to which to compare the performance of $ACO_{\mathbb{R}}$. In order to ensure a fair comparison, we have re-implemented some algorithms traditionally used for training NNs—namely the back-propagation (BP) algorithm and the Levenberg-Marquardt (LM) algorithm. We used the R programming language (a free alternative to S+) for implementing these algorithms.[7]

Back-Propagation is a gradient-descent algorithm traditionally used for training NNs (Rumelhart *et al.*, 1986). It is a first-order minimization algorithm—i.e. it is based on first-order derivatives (i.e., the gradient). It uses the estimation of the gradient of the instantaneous sum-squared error for each network layer:

$$\Delta \vec{w} = -\eta \nabla E(\vec{w}) \quad , \tag{10}$$

where $\vec{w}$ is the vector of all weights, η is the learning rate, and E is the gradient. The algorithm we have implemented is the basic version of back-propagation without heuristic improvements that were developed over time.

Levenberg-Marquardt is a variation of Newton's method that was initially designed for minimizing functions that are either sums of squares,

[7] http://www.R-project.org

or, in general, other non-linear functions (Hagan and Menhaj, 1994; Papliński, 2004). In Newton's method, minimization is based on utilizing the second order derivatives as well as on the use of a batch training mode rather than the pattern mode (which is used, for example, in back-propagation). The batch training mode is based on derivatives of instantaneous errors. The LM algorithm uses an approximation of the Hessian matrix by adding a small constant μ multiplied by the identity matrix I to the product of the transposed Jacobian matrix J^T and the Jacobian matrix J:

$$\Delta\vec{w} = -\sum_{1}^{P} \nabla E(w)\left[J(w)^T J(w) + \mu I\right]^{-1} \tag{11}$$

Both algorithms (i.e., BP and LM) require gradient information. Hence, they require the neuron transfer function to be differentiable. Consequently, these algorithms may not be used in case, when the neuron transfer function is not differentiable or is unknown. In contrast, $ACO_{\mathbb{R}}$ is a general heuristic optimization that can be applied when the neuron transfer function is non-differentiable. On the other side, in case, when the neuron transfer function is differentiable, the drawback of general optimization algorithms such as $ACO_{\mathbb{R}}$ is that they do not exploit available additional information as, for example, gradient information.

In order to see how the additional gradient information influences the performance of $ACO_{\mathbb{R}}$, we have also implemented hybridized versions of $ACO_{\mathbb{R}}$, namely $ACO_{\mathbb{R}}$-BP and $ACO_{\mathbb{R}}$-LM, which are hybrids of the $ACO_{\mathbb{R}}$ algorithm and respectively the BP and LM algorithms. In these hybrids, each solution generated by the $ACO_{\mathbb{R}}$ algorithm is improved by running a single improving iteration of either BP or LM, respectively.

Finally, we wanted to study how all the algorithms tested compare to a simple random restart search method. In order to accomplish that, we have implemented random search (RS)—i.e. an algorithm that randomly generates a set of values for the weights and then evaluates these solutions. As we used a sigmoid function as neuron transfer function, it was sufficient to limit the range of weight values to values close to 0. Hence, we arbitrarily chose a range of [-5,5].

4.4 Parameter Tuning

All our algorithms (with the exception of RS) require certain parameter values to be determined before they can be applied. While algorithms such as BP or LM have very few parameters, $ACO_{\mathbb{R}}$, as well as its hybridized versions, have more. In general, in order to ensure a fair comparison of algorithms, an equal amount of effort is required in the parameter tuning process for each of the algorithms. Also, it has been shown in the literature that the stopping condition for the parameter tuning runs should be identical to the one used in the actual experiments (be that time, number of iterations, etc.), as otherwise the danger of choosing suboptimal parameter values increases (Socha, 2003). We have hence used a common parameter tuning methodology for all our algorithms, with the same stopping condition that we planned to use for the final experiments. The methodology that we used is known as F-RACE methodology (Birattari *et al.*, 2002; Birattari, 2004). In particular we used the RACE package[8] for R. It allows running a *race* of different configurations of algorithms against each other on a set of test instances. After each round, the non-parametric Friedman test is used to compare the performance of different configurations. Configurations are being dropped from the race as soon as sufficient statistical evidence has been gathered against them. For more information on the F-RACE methodology, we refer the interested reader to (Birattari, 2004). Since for the problems we investigated we did not have several instances available (i.e., we wanted to tune the algorithms for each of the three considered data sets separately), we have created a set of instances for each race by dividing randomly (several times) the training set of each problem instance into a training set for tuning (two thirds of the training set) and a test set for tuning (one third of the training set). Table 8-2 provides details on the number of patterns used respectively for learning and validation during the parameter tuning runs, as well as for training and testing the chosen configurations.

Table 8-2. Summary of the number of patterns used for training and testing, both for parameter tuning as well as for the final performance evaluation. The patterns used for parameter tuning (learning and testing) were randomly chosen from the training set that we used later in the performance evaluation

Algorithm	*Total number of patterns*	*Parameter tuning*		*Performance evaluation*	
		Training set for tuning	*Test set for tuning*	*Training Set*	*Test Set*
Cancer1	699	350	175	525	174
Diabetes1	768	384	192	576	192
Heart1	920	460	230	690	230

For the tuning, we determined 10 different configurations of parameter settings for each of our algorithms. Then, we applied the F-RACE to each instance set (i.e., per algorithm, per problem), allowing not more than 100

[8] http://cran.r-project.org/src/contrib/Descriptions/race.html

experiments in the race. Each of the parameter tuning races returned one configuration that performed best[9]. The final parameter value settings that we used for our final experiments are summarized in Table 8-3.

Table 8-3. Summary of the parameters chosen for our algorithms. Not included in the table are the parameters common to all ACO_R versions, namely q and m. For these parameters we used the settings q=0.01, and m=2 (the number of ants used in each iteration)

	Cancer1				Diabetes1				Heart1			
Algorithm	k	ρ	η	β	k	ρ	η	β	k	ρ	η	β
ACO_R	148	0.95	-	-	136	0.8	-	-	230	0.6	-	-
ACO_R -BP	148	0.98	0.3	-	136	0.7	0.1	-	230	0.98	0.4	-
ACO_R -LM	148	0.9	-	10	136	0.1	-	10	230	0.1	-	10
BP	-	-	0.002	-	-	-	0.01	-	-	-	0.001	-
LM	-	-	-	50	-	-	-	5	-	-	-	1.5

4.5 Results

In order to compare the performance of the algorithms, we applied each algorithm 50 times to each of the three test problems. As stopping condition we used the number of fitness function evaluations. Following the work of Alba and Chicano (2004), we used 1000 function evaluations as the limit. We used the training and testing approach—no cross-validation.

Figures 8-7, 8-8, and 8-9 present respectively the results obtained for the cancer, diabetes, and heart test problems in the form of box-plots. Each figure contains two graphics; the left one presents the distributions of the actual CEP values obtained by the algorithms (over 50 independent runs); the right one presents the distributions of rankings achiseved by the algorithms. Any solution generated by any of the algorithms is ranked. Having 6 algorithms and running 50 trials each, the possible rankings vary from 1 to 300. The distribution of those rankings is then plotted per algorithm—this allows for a clear identification of those better performing ones, regardless of how small the difference may be in terms of objective function value. The boxes are drawn between the first and the third quartile of the distribution, while the indentations in the box-plots (or *notches*) indicate the 95% confidence interval for a given distribution (McGill et. al, 1978). In other words, this means that if the notches of two distributions do not overlap, they are significantly different with 95% confidence.

[9] Due to the limited resources for tuning, the chosen configuration for each race is not necessarily significantly better than all the others. The limit of 100 experiments per race did sometimes not allow reaching that level of assurance. However, the chosen configuration was definitely not significantly worse than any of the others.

Cancer (Fig. 8-7) appears to be the easiest problem among the three. All algorithms obtained reasonably good results, including the RS method (!). However, the best performing algorithm is BP. From the fact that the results obtained by RS do not differ significantly from the results obtained by other—more complex—algorithms, it may be concluded that the problem is relatively easy, and that there are a lot of reasonably good solutions scattered over the search space. None of the algorithms was able to classify all the test patterns correctly.

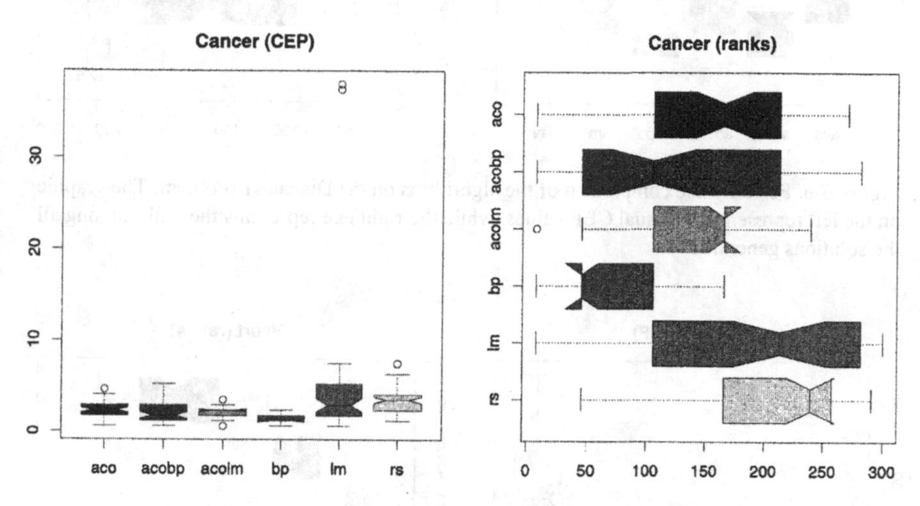

Figure 8-7. Performance comparison of the algorithms on the Cancer1 problem. The graphic on the left represents the actual CEP values, while the right one represents the ranks among all the solutions generated.

Diabetes (Fig. 8-8) is a problem that is more difficult than Cancer. All our algorithms clearly outperform RS. However, the overall performance of the algorithms is not very good. The best performing is again BP. The less good overall performance of the algorithms may again indicate that the training set does not represent fully all the possible patterns.

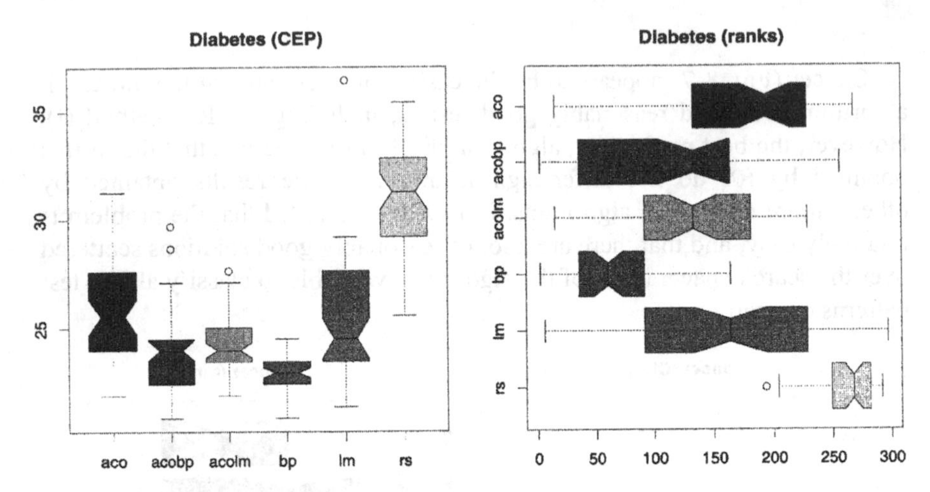

Figure 8-8. Performance comparison of the algorithms on the Diabetes1 problem. The graphic on the left represents the actual CEP values, while the right one represents the ranks among all the solutions generated.

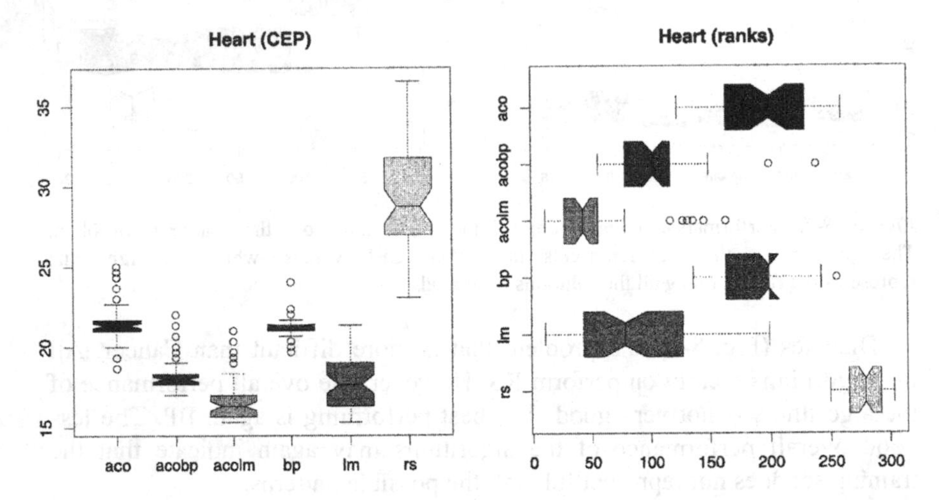

Figure 8-9. Performance comparison of the algorithms on the Heart1 problem. The graphic on the left represents the actual CEP values, while the right one represents the ranks among all the solutions generated.

The Heart problem (Fig. 8-9) with 230 weight values is the largest problem that we tackled. It is also the one on which the performance of the algorithms differed mostly. All tested algorithms clearly outperform RS, but there are also significant differences among the more complex algorithms. BP, which was performing quite well on the other two test problems, did not do so well on Heart. ACO_R achieves results similar to BP. In turn, LM, which was not performing so well on the first two problems, obtains quite good results. Very interesting is the performance of the hybridized versions of ACO_R — ACO_R-BP and ACO_R -LM. The ACO_R-BP hybrid clearly outperforms both ACO_R and BP. ACO_R-LM outperforms respectively ACO_R and LM. Additionally, ACO_R-LM performs best overall.

Summarizing, we note that the performance of ACO_R alone does often not quite reach the performance of the derivative based algorithms and the ACO_R hybrids. Its performance is, however, not much worse. Furthermore, the results show that hybridizing ACO_R with BP or LM helps to improve the results of the pure ACO_R algorithm. This was especially the case for Heart, where ACO_R-LM was the overall winner. We want to remind at this point that ACO_R is much more general than for example BP and LM, because it does not require derivative information. Hence, it may be applied when the neuron transfer function of a NN is non-differentiable or unknown, while algorithms such as BP or LM could not be used in this case.

Table 8-4. Pair-wise comparison of the results of ACO_R-based algorithms with recent results obtained by a set of GA based algorithms (Alba and Chicano, 2004). The results can be compared thanks to maintaining the same experimental setup. For each problem-algorithm pair we give the mean (over 50 independent runs), and the standard deviation (in brackets). The best result of each comparison is indicated in bold

	GA	*ACO_R*	*GA-BP*	*ACO_R-BP*	*GA-LM*	*ACO_R-LM*
Cancer	16.76 (6.15)	**2.39 (1.15)**	**1.43 (4.87)**	2.14 (1.09)	**0.02 (0.11)**	2.08 (0.68)
Diabetes	36.46 (0.00)	**25.82 (2.59)**	36.46 (0.00)	**23.80 (1.73)**	28.29 (1.15)	**24.26 (1.40)**
Heart	41.50 (14.68)	**21.59 (1.14)**	54.30 (20.03)	**18.29 (1.00)**	22.66 (0.82)	**16.53 (1.37)**

Finally it is interesting to compare the performance of the ACO_R based algorithms to some other general optimization algorithms. Alba and Chicano (2004) have published the results of a Genetic Algorithm (GA) used for tackling exactly the same three problems as we did. They have tested not only a stand-alone GA, but also its hybridized versions: GA-BP and GA-LM.

Table 8-4 summarizes the results obtained by the ACO_R and GA based algorithms. Clearly the stand-alone ACO_R performs better than the stand-alone GA for all the test problems. ACO_R-BP and ACO_R-LM perform respectively better than GA-BP and GA-LM on both of the more difficult problems—Diabetes and Heart—and worse on Cancer. For the Heart problem the mean performance of *any* $ACO_{\mathbb{R}}$ based algorithm is significantly better than the *best* GA based algorithm (which was reported as the state-of-the-art for this problem in 2004).

5. CONCLUSIONS

We have presented an ant colony optimization algorithm (i.e., $ACO_{\mathbb{R}}$) for the training of feed-forward neural networks in classification problems. $ACO_{\mathbb{R}}$ is a generic approach that can be flexibly used either as a stand-alone method, or hybridized with more problem specific algorithms. The performance of the algorithm was evaluated on real-world test problems and compared to specialized algorithms for feed-forward neural network training (back propagation and Levenberg-Marquardt), and also to genetic algorithm based algorithms.

The performance of the stand-alone $ACO_{\mathbb{R}}$ was comparable (or at least not much worse) than the performance of specialized algorithms for neural network training. This result is particularly interesting as $ACO_{\mathbb{R}}$—being a much more generic approach—allows also the training of networks in which the neuron transfer function is either not differentiable or unknown. The hybrid between $ACO_{\mathbb{R}}$ and the Levenberg-Marquardt algorithm (i.e., $ACO_{\mathbb{R}}$-LM) was in some cases able to outperform the back propagation and the Levenberg-Marquardt algorithms that are traditionally used for neural network training. Finally, when compared to other general-purpose algorithms, namely genetic algorithm based algorithms from the literature, our results showed that the ant colony optimization based algorithms may provide superior performance for some of the test problems.

ACKNOWLEDGEMENTS

This work was supported by the Spanish CICYT project no. TIC-2002-04498-C05-03 (TRACER), and by the "Juan de la Cierva" program of the Spanish Ministry of Science and Technology of which Christian Blum is a post-doctoral research fellow.

This work was also partially supported by the "ANTS" project, an "Action de Recherche Concertée" funded by the Scientific Research Directorate of the French Community of Belgium.

REFERENCES

Alba, E., and Chicano, J.F, 2004, Training Neural Networks with GA Hybrid Algorithms, in: *Proceedings of Genetic and Evolutionary Computation - GECCO 2004, Part 1*, Lecture Notes in Computer Science, vol. 3102, K. Deb et al, eds., Springer-Verlag, Berlin, Germany, pp. 852-863.

Battiti, R., and Tecchiolli, G., 1996, The continuous reactive tabu search: Blending combinatorial optimization and stochastic search for global optimization, *Annals of Operations Research* **63**:153-188.

Bilchev, G., and Parmee, I. C., 1995, The ant colony metaphor for searching continuous design spaces, in: *Proceedings of the AISB Workshop on Evolutionary Computation*, Lecture Notes in Computer Science, vol. 993, T.~C. Fogarty, ed., Springer-Verlag, Berlin, Germany, pp. 25-39.

Birattari, M., 2004, *The Problem of Tuning Metaheuristics as Seen from a Machine Learning Perspective*, Ph.D. thesis, ULB, Brussels, Belgium.

Birattari, M., Stützle, T., Paquete, L., and Varrentrapp, K., 2002, A Racing Algorithm for Configuring Metaheuristics, in: *Proceedings of Genetic and Evolutionary Conference*, W. B. Langdon *et al.* eds., Morgan Kaufmann, San Francisco, CA, USA, pp. 11-18.

Blum, C., 2005, Beam-ACO—Hybridizing ant colony optimization with beam search: An application to open shop scheduling, *Computers & Operations Research* **32**(6):1565-1591.

Blum, C., and Roli, A., 2003, Metaheuristics in combinatorial optimization: Overview and conceptual comparison, *ACM Computing Surveys* **35**(3):268-308.

Blum, C., and Sampels, M., 2004, An ant colony optimization algorithm for shop scheduling problems, *Journal of Mathematical Modelling and Algorithms* **3**(3):285-308.

Blum, C., 2005, Beam-ACO—Hybridizing ant colony optimization with beam search: An application to open shop scheduling, *Computers & Operations Research* **32**(6):1565-1591.

Bonabeau, E., Dorigo, M., and Theraulaz, G., 1999, *Swarm Intelligence: From Natural to Artificial Systems*, Oxford University Press, New York, NY.

Box, G. E. P., and Muller, M. E, 1958, A note on the generation of random normal deviates. *Annals of Mathematical Statistics* **29**(2):610-611.

Černý, V., 1985, A thermodynamical approach to the travelling salesman problem: An efficient simulation algorithm, *Optimization Theory and Applications* **45**:41-51.

Chelouah, R., and Siarry, P., 2000, A continuous genetic algorithm designed for the global optimization of mulitmodal functions, *Journal of Heuristics* **6**:191-213.

Chelouah, R., and Siarry, P., 2000, Tabu search applied to global optimization, *European Journal of Operational Research* **123**:256-270.

Chelouah, R., and Siarry, P., 2003, Genetic and Nelder-Mead algorithms hybridized for a more accurate global optimization of continuous multiminima functions, *European Journal of Operational Research* **148**:335-348.

Costa, D., and Hertz, A., 1997, Ants can color graphs, *Journal of the Operational Research Society* **48**:295-305.

den Besten, M. L., Stützle, T., and Dorigo, M., 2000, Ant colony optimization for the total weighted tardiness problem, in: *Proceedings of PPSN-VI, Sixth International Conference on Parallel Problem Solving from Nature*, Lecture Notes in Computer Science, vol. 1917, M.~Schoenauer et al., eds., Springer Verlag, Berlin, Germany, pp. 611-620.

Deneubourg, J.-L., Aron, S., Goss, S., and Pasteels, J.-M., 1990, The self-organizing exploratory pattern of the argentine ant, *Journal of Insect Behaviour* **3**:159-168.

Dorigo, M., 1992, *Optimization, Learning and Natural Algorithms* (in Italian), PhD thesis, Dipartimento di Elettronica, Politecnico di Milano, Italy.

Dorigo, M., and Gambardella, L. M., 1997, Ant Colony System: A cooperative learning approach to the travelling salesman problem, *IEEE Transactions on Evolutionary Computation* **1**(1):53-66.

Dorigo, M., Maniezzo, V., and Colorni, A., 1991, Positive feedback as a search strategy, Technical Report 91-016, Dipartimento di Elettronica, Politecnico di Milano, Italy.

Dorigo, M., Maniezzo, V., and Colorni, A., 1996, Ant System: Optimization by a colony of cooperating agents, *IEEE Transactions on Systems, Man, and Cybernetics — Part B* **26**(1):29-41.

Dorigo, M., and Stützle, T., 2004, *Ant Colony Optimization*, MIT Press, Cambridge, MA.

Dréo, J., and Siarry, P., 2002, A new ant colony algorithm using the heterarchical concept aimed at optimization of multiminima continuous functions, in: *Proceedings of ANTS 2002—From Ant Colonies to Artificial Ants: Third International Workshop on Ant Algorithms*, Lecture Notes in Computer Science, vol. 2463 of LNCS, M. Dorigo et al., eds., Springer Verlag, Berlin, Germany, pp. 216-221.

Fogel, L. J., Owens, A. J., and Walsh, M. J., 1966, *Artificial Intelligence through Simulated Evolution*, Wiley.

Gagné, C., Price, W. L., and Gravel, M., 2002, Comparing an ACO algorithm with other heuristics for the single machine scheduling problem with sequence-dependent setup times, *Journal of the Operational Research Society* **53**:895-906.

Gambardella, L. M., and Dorigo, M., 2000, Ant Colony System hybridized with a new local search for the sequential ordering problem, *INFORMS Journal on Computing* **12**(3):237-255.

Gambardella, L. M., Taillard, É. D., and Agazzi, G., 1999, MACS-VRPTW: A multiple ant colony system for vehicle routing problems with time windows, in: *New Ideas in Optimization*, D. Corne et al., eds., McGraw Hill, London, UK, pp. 63-76.

Glover, F., 1989, Tabu search—Part I, *ORSA Journal on Computing* **1**(3):190-206.

Glover, F., 1990, Tabu search—Part II, *ORSA Journal on Computing* **2**(1):4-32.

Glover, F., and Kochenberger, G., 2002, *Handbook of Metaheuristics*, Kluwer Academic Publishers, Norwell, MA.

Glover, F., and Laguna, M., 1997, *Tabu Search*, Kluwer Academic Publishers.

Goldberg, D. E., 1989, *Genetic algorithms in search, optimization, and machine learning*, Addison Wesley, Reading, MA.

Golub, G. H., and van Loan, C. F., 1989, *Matrix Computations*, 2nd ed., the John Hopkins University Press, Baltimore, MD, USA.

Guntsch, M., and Middendorf, M., 2002, A population based approach for ACO, in: *Applications of Evolutionary Computing, Proceedings of EvoWorkshops 2002: EvoCOP, EvoIASP, EvoSTim*, vol. 2279, S. Cagnoni, J. Gottlieb, E. Hart, M. Middendorf, and G. Raidl, eds., Springer-Verlag, Berlin, Germany, pp. 71-80.

Hagan, M. T., and Menhaj, M. B., 1994, Training Feedforward Networks with the Marquardt Algorithm, *IEEE Transactions on Neural Networks* **5**:989-993.

Hastie, T., Tibshirani, R., and Friedman, J., 2001, *The Elements of Statistical Learning*, Springer-Verlag, Berlin, Germany.

Holland, J. H., 1975, *Adaption in natural and artificial systems*, The University of Michigan Press, Ann Harbor, MI.

Hoos, H. H., and Stützle, T., 2004, *Stochastic Local Search: Foundations and Applications*, Elsevier, Amsterdam, The Netherlands.

Kern, S., Müller, S. D., Hansen, N., Büche, D., Očenášek, J., and Koumoutsakos, P., 2004, Learning probability distributions in continuous evolutionary algorithms—A comparative review, *Natural Computing* **3**(1):77-112.

Kirkpatrick, S., Gelatt, C. D., and Vecchi, M. P., 1983, Optimization by simulated annealing, *Science* **220**(4598):671-680.

Maniezzo, V., 1999, Exact and Approximate Nondeterministic Tree-Search Procedures for the Quadratic Assignment Problem, *INFORMS Journal on Computing* **11**(4):358-369.

Maniezzo, V., and Colorni, A., 1999, The Ant System applied to the quadratic assignment problem, *IEEE Transactions on Data and Knowledge Engineering* **11**(5):769-778.

Mathur, M., Karale, S. B., Priye, S., Jyaraman, and V. K., Kulkarni, B. D., 2000, Ant colony approach to continuous function optimization, *Industrial & Engineering Chemistry Research* **39**:3814-3822.

McGill, R., Tukey, J. W.,Larsen, and W. A., 1978, Variations of box plots, *The American Statisticia* **32**:12-16 .

Merkle, D., Middendorf, M., and Schmeck, H., 2002, Ant Colony Optimization for Resource-Constrained Project Scheduling, *IEEE Transactions on Evolutionary Computation* **6**(4):333-346.

Monmarché, N., Venturini, and G.,Slimane M., 2000, On how Pachycondyla apicalis ants suggest a new search algorithm, *Future Generation Computer Systems* **16**:937-946.

Nelder, J. A., and Mead, R., 1965, A simplex method for function minimization, *Computer Journal* **7**:308-313.

Papadimitriou, C. H., and Steiglitz, K., 1982, *Combinatorial Optimization—Algorithms and Complexity*, Dover Publications, Inc., New York.

Papliński, A.P., 2004, Lecture 7—Advanced Learning Algorithms for Multilayer Perceptrons, available online at http://www.csse.monash.edu.au/courseware/cse5301/04/L07.pdf.

Prechelt, L., 1994, Proben1—A Set of Neural Network Benchmark Problems and Benchmarking Rules. Technical Report 21, Fakultät für Informatik, Universität Karlsruhe, Karlsruhe, Germany.

Rechenberg, I., 1973, *Evolutionsstrategie: Optimierung technischer Systeme nach Prinzipien der biologischen Evolution*, Frommann-Holzboog.

Reimann, M., Doerner, K., and Hartl, R. F., 2004, D-ants: Savings based ants divide and conquer the vehicle routing problems, *Computers & Operations Research* **31**(4):563-591.

Rumelhart, D., Hinton, G., and Williams, R., 1986, Learning Representations by Backpropagation Errors, *Nature* **323**:533-536.

Siarry, P., Berthiau, G., Durbin, F., and Haussy, J., 1997, Enhanced simulated annealing for globally minimizing functions of many-continuous variables, *ACM Transactions on Mathematical Software* **23**(2):209.228.

Socha, K., 2003, The Influence of Run-Time Limits on Choosing Ant System Parameters, in *Proceedings of GECCO 2003—Genetic and Evolutionary Computation Conference*, Lecture Notes in Computer Science, vol. 2723, E. Cantu-Paz et al., eds., Springer-Verlag, Berlin, Germany, pp. 49-60.

Socha, K., 2004, Extended ACO for continuous and mixed-variable optimization, in: *Proceedings of ANTS 2004—Fourth International Workshop on Ant Algorithms and Swarm Intelligence*, Lecture Notes in Computer Science, M. Dorigo et al., eds., Springer Verlag, Berlin, Germany, pp. 35-46.

Socha, K., Sampels, M., and Manfrin, M., 2003, Ant algorithms for the university course timetabling problem with regard to the state-of-the-art, in: *Applications of Evolutionary Computing, Proceedings of EvoWorkshops 2003*, vol. 2611, G. Raidl et al., eds., pp 334-345.

Storn, R., and Price, K., 1997, Differential evolution—A simple and efficient heuristic for global optimization over continuous spaces, *Journal of Global Optimization* **11**:341-359.

Stützle, T., 1998, An Ant Approach to the Flow Shop Problem, in: *Proceedings of the Fifth European Congress on Intelligent Techniques and Soft Computing, EUFIT'98*, pp 1560-1564.

Stützle, T., and Hoos, H. H., 2000, *MAX-MIN* Ant System, *Future Generation Computer Systems* **16**(8):889-914.

Chapter 9

COOPERATIVE COEVOLUTIONARY METHODS

Nicolás García-Pedrajas, César Hervás-Martínez and Domingo Ortiz-Boyer
Department of Computing and Numerical Analysis, University of Córdoba, Spain

Abstract: This chapter presents a cooperative coevolutionary model for evolving artificial neural networks. This model is based on the idea of coevolving subnetworks that must cooperate to form a solution for a specific problem, instead of evolving complete networks. The combination of these subnetworks is part of a coevolutionary process. The best combinations of subnetworks must be evolved together with the coevolution of the subnetworks. Several subpopulations of subnetworks coevolve cooperatively and genetically isolated. The individuals of every subpopulation are combined to form whole networks. This is a different approach from most current models of evolutionary neural networks which try to develop whole networks. This model places as few restrictions as possible over the network structure, allowing the model to reach a wide variety of architectures during the evolution and to be easily extensible to other kind of neural networks. The performance of the model in solving ten real problems of classification is compared with a modular network, the adaptive mixture of experts, and with the results reported in the literature.

Key words: Neural network automatic design; cooperative coevolution; evolutionary computation; genetic algorithms; evolutionary programming.

1. INTRODUCTION

In the area of neural networks (Haykin, 1994) design one of the main problems is finding suitable architectures for solving specific tasks. The election of such architecture is very important, as a network smaller than needed would be unable to learn and a network larger than needed would end in over-training.

The problem of finding a suitable architecture and the corresponding weights of the network is a very complex task (Yao, 1999). Modular systems are often used in machine learning as an approach for solving those complex problems. Moreover, in spite of the fact that small networks are preferred because they usually lead to better performance, the error surfaces of such networks are more rugged and have few good solutions (Shang and Wah, 1996). In addition, there is much neuropsychological evidence showing that the brain of humans and other animals consists of modules, which are subdivisions in identifiable parts, each one with its own purpose and function (Cho and Shimohara, 1998).

The objective of this chapter is showing how Cooperative Coevolution, can be used to design such modular neural networks. Evolutionary computation (Goldberg, 1989c; Michalewicz, 1994) is a set of global optimization techniques that have been widely used for training and automatically designing neural networks (see Section 3). Some efforts have been made in designing modular (Caelli et al., 1999) neural networks with these techniques (Yao and Liu, 1997), but in almost all of them the design of the networks is helped by methods outside evolutionary computation, or the application area for those models is limited to very specific architectures.

This chapter is organised as follows: Section 2 explains the paradigm of cooperative coevolution; Section 3 shows an application of cooperative coevolution to automatic neural network design; Section 4 describes the experiments carried out; and finally Section 5 states the conclusions of this chapter.

2. COOPERATIVE COEVOLUTION

Cooperative coevolution (Potter, 1997) is a recent paradigm in the area of evolutionary computation focused on the evolution of coadapted subcomponents without external interaction. In cooperative coevolution a number of species are evolved together. The cooperation among the individuals is encouraged by rewarding the individuals based on how well they cooperate to solve a target problem. The work on this paradigm has shown that cooperative coevolutionary models present many interesting features, such as specialization through genetic isolation, generalization and efficiency (Potter and de Jong, 2000). Cooperative coevolution approaches the design of modular systems in a natural way, as modularity is part of the model. Other models need some a priori knowledge to decompose the problem by hand. In many cases, either this knowledge is not available or it is not clear how to decompose the problem.

This chapter describes a cooperative coevolutionary model called Covnet (García-Pedrajas et al., 2003). This model develops subnetworks instead of whole networks. These modules are combined to form a network. As M. A. Potter and K. A. De Jong (2000) have stated, "*to apply evolutionary algorithms effectively to increasingly complex problems explicit notions of modularity must be introduced to provide reasonable opportunities for solutions to evolve in the form of interacting coadapted subcomponents*".

The most distinctive feature of Covnet is the coevolution of modules without the intervention of any agent external to the evolutionary process and without an external mechanism for combining subnetworks. Also, the use of an evolutionary algorithm for the evolution of both the weights and the architecture allows the model to be applied to tasks where there is no error function that could be defined (e.g.: game playing (Chellapilla and Fogel, 1999) or control (Lin and Jou, 1999)) in order to apply an algorithm based on the minimization of that error, like the backpropagation learning rule, or the derivatives of that error function cannot be obtained.

The most important contribution of Covnet are the following. First, it forms modular artificial neural networks using cooperative coevolution. Every module must learn how to combine with the other modules of the network to be useful. Introducing the combination of modules into the evolutionary process enforces the cooperation among the modules, as independently evolved modules are less likely to combine well after the evolutionary process have finished.

Second, it develops a method for measuring the fitness of cooperative subcomponents in a coevolutionary model. This method, based on three different criteria, could be applied to other cooperative coevolutionary models not related to the evolution of neural networks. The current methods are based, almost exclusively, on measuring the fitness of the networks where the module appears.

Third, it introduces a new hybrid evolutionary programming algorithm that puts very few restrictions in the subnetworks evolved. This algorithm produces very compact subnetworks, and even the evolved subnetworks alone achieved very good performance in the test problems.

3. AUTOMATIC DESIGN OF ARTIFICIAL NEURAL NETWORKS BY MEANS OF COOPERATIVE COEVOLUTION

The automatic design of artificial neural networks has two different approaches: parametric learning and structural learning. In structural learning, both architecture and parametric information must be learned through the process of training. Basically, we can consider three models of structural learning: constructive algorithms, destructive algorithms, and evolutionary computation.

Constructive algorithms (Gallant, 1993; Honavar and Uhr, 1993; Parekh et al., 2000) start with a small network (usually a single neuron). This network is trained until it is unable to continue learning, then new components are added to the network. This process is repeated until a satisfactory solution is found. These methods are usually trapped in local minima (Angeline et al., 1994) and tend to produce large networks. Destructive methods, also known as pruning algorithms (Reed, 1993), start with a big network, that is able to learn but usually ends in over-fitting, and try to remove the connections and nodes that are not useful. A major problem with pruning methods is measuring the relevance of the structural components of the network in order to decide whether a connection or node must be removed.

Both methods, constructive and destructive, limit the number of available architectures thus introducing constraints in the search space of possible structures that may not be suitable to the problem. Although these methods have been proved useful in simulated data (Depenau and Moller, 1994; Thodberg, 1991), their application to real problems has been rather unsuccessful (Hirose et al., 1991; Hassibi and Stork, 1993; Kamimura and Nakanishi, 1994).

Evolutionary computation has been widely used to evolve neural network architectures and weights. There have been many applications for parametric learning (van Rooij et al., 1996) and for both parametric and structural learning (Yao and Liu, 1997; Angeline et al., 1994; Odri et al., 1993; Smalz and Conrad, 1994; Maniezzo, 1994; Borst, 1994; Whitehead and Choate, 1996; Moriarty, 1997). These works fall into two broad categories of evolutionary computation: genetic algorithms and evolutionary programming.

Genetic algorithms are based on a representation independent of the problem, usually the representation is a string of binary, integer or real numbers. This representation (the genotype) codifies a network (the phenotype). This is a dual representation scheme. The ability to create better solutions in a genetic algorithm relies mainly on the crossover operator. This operator create new individuals by recombining representational components from two members of the population.

The benefits of crossover come from the ability of forming connected substrings of the representation that correspond to above-average solutions (Goldberg, 1989c). These substrings are called *building blocks*. Crossover is not effective in environments where the fitness of an individual of the population is not correlated with the expected ability of its representational components (Goldberg, 1989b). Such environments are called *deceptive* (Goldberg, 1989a). Deception is a very important feature in most representations of neural networks, so crossover is usually avoided in evolutionary neural networks (Angeline et al., 1994).

One of the most important forms of deception arises from the many-to-one mapping from genotypes in the representation space to phenotypes in the evaluation space. The existence of networks functionally equivalent and with different encodings makes the evolution inefficient, and it is unclear whether crossover would produce more fitted individuals from two members of the population. This problem is usually termed as the *permutation problem* (Belew et al., 1991; Hancock, 1992) or the *competing conventions problem* (Schaffer et al., 1992).

Evolutionary programming (Fogel, 1992) is, for many authors, the most suited paradigm of evolutionary computation for evolving artificial neural networks (Angeline et al., 1994). Evolutionary programming uses a representation natural for the problem. Once the representation scheme has been chosen, mutation operators specific to the representation scheme are defined. Evolutionary programming offers a major advantage over genetic algorithms when evolving artificial neural networks, the representation scheme allows manipulating networks directly, avoiding the problems associated with a dual representation.

The use of evolutionary learning for designing neural networks dates from no more than two decades (Yao, 1999; Schaffer et al., 1992). However, a lot of work has been made in these two decades, with many different approaches and working models (Yao and Liu, 1997; Odri et al., 1993; Schaffer et al., 1992). Evolutionary computation has been used for learning connection weights and for learning both architecture and connection weights. The main advantage of evolutionary computation is that it performs a global exploration of the search space avoiding to become trapped in local minima as usually happens with local search procedures.

G. F. Miller et al. (1991) proposed that evolutionary computation is a very good candidate to be used to search the space of topologies because the fitness function associated with that space is complex, noisy, non-differentiable, multi-modal and deceptive.

Almost all the current models try to develop a global architecture, which is a very complex problem. Although, some attempts have been made in developing modular networks (Liu and Yao, 1999; Rosen, 1996), in most cases the modules are combined only after the evolutionary process has finished and not following a cooperative coevolutionary model.

Few authors have devoted their attention to the cooperative coevolution of subnetworks. Some authors have termed this kind of cooperative evolution (where the individuals must cooperate to achieve a good performance) *symbiotic evolution* (Moriarty and Miikkulainen, 1996). More formally, we should speak of mutualism, that is, the cooperation of two individuals from different species that benefits both organisms.

R. Smalz and M. Conrad (1994) developed a cooperative model where there are two populations: a population of nodes, divided into clusters, and a population of networks that are combinations of neurons, one from each cluster. Both populations are evolved separately.

B. A. Whitehead and T. D. Choate (1996) developed a cooperative-competitive genetic model for Radial-Basis Function (RBF) neural networks. In this work there is a population of genetically encoded neurons that evolves both the centres and the widths of the radial basis functions. There is just one network that is formed by the whole population of RBF's. The major problem, as in our approach, is to obtain the fitness of each node of the population, as the only performance measure available is for the whole network. This is well known as the "*credit apportionment problem*" (Samuel, 1959; Smalz and Conrad, 1994; Potter, 1997). The credit assignment used by Whitehead and Choate is restricted to RBF-like networks and very difficult to adapt to other kind of networks.

D. W. Opitz and J. W. Shavlik (1996) developed a model called ADDEMUP (*Accurate anD Diverse Ensemble Maker giving United Predictions*). They evolved a population of networks by means of a genetic algorithm and combined the networks in an ensemble. The competition among the networks is encouraged with a diversity term added to the fitness of each network.

D. E. Moriarty and R. Miikkulainen (1996; Moriarty, 1997) developed an actual cooperative model, called SANE, that had some common points with the work of R. Smalz and M. Conrad (1994). In this work they propose two populations: one of nodes and another of networks that are combinations of the individuals from the population of nodes. Zhao et al. (2000) proposed a framework for cooperative coevolution, and applied that framework to the evolution of RBF networks. Nevertheless, their work, more than a finished model, is an open proposal that aims at the definition of the problems to be solved in a cooperative environment.

S-B. Cho and K. Shimohara (1998) developed a modular neural network evolved by means of genetic programming. Each network is a complex structure formed by different modules which are codified by a tree structure.

X. Yao and Y. Liu (1998) use the final population of networks developed using EPNet (Yao and Liu, 1997) model to form ensembles of neural networks. The combination of these networks produced better results than any isolated network. Nevertheless, the cooperation among the networks takes place only after the evolutionary process has finished.

3.1 Covnet: A Cooperative Coevolutionary Model

Covnet is a cooperative coevolutionary model, that is, several species are coevolved together. Each species is a subnetwork that constitutes a partial solution of a problem; the combination of several individuals from different species constitutes the network. The population of subnetworks, that are called *nodules*, is made up by several subpopulations. Each subpopulation evolves independently, so we can talk of subpopulations or species indistinctly, as each subpopulation will constitute a different species. The combination of individuals from these different subpopulations is the key factor of Covnet.

The evolution of coadapted subcomponents must address four major issues: problem decomposition, interdependence among subcomponents, credit assignment and maintenance of diversity. Cooperative coevolution gives a framework where these issues could be faced in a natural way. The problem decomposition is intrinsic in the model. Each population will evolve different species that must cooperate in order to be rewarded with high fitness values. There is no need of any a priori knowledge to decompose the problem by hand. The interdependence among the subcomponents comes from the fact that the fitness of each individual depends on how well the individual works together with the members of other species.

A nodule is made up of a variable number of nodes with free interconnection among them (see Figure 9-1), that is, each node may have connections from input nodes, from other nodes of the nodule, and to output nodes. More formally a nodule could be defined as follows:

Definition 1: (Nodule) A nodule is a subnetwork formed by: a set of nodes with free interconnection among them, the connection of these nodes from the input and the connections of the nodes to the output. It cannot have connections with any node belonging to another nodule.

The input and output layers of the nodules are common, they are the input and output layers of the network. It is important to note that the genotype of the nodule has a one-to-one mapping to the phenotype, as the many-to-one mapping between them is one of the main sources of deception and the permutation problem (Angeline et al., 1994).

In the same way we define a network as a combination of nodules. The definition more formally is as follows:

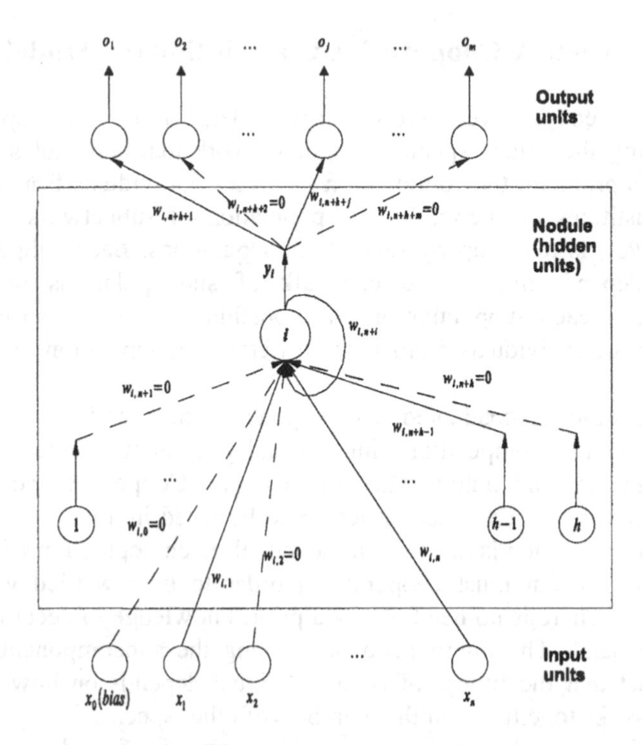

Figure 9-1. Model of a nodule. As a node has only connections to some nodes of the nodule, the connections that are missing are represented with dashed lines. The nodule is composed by the hidden nodes and the connections of these nodes from the input and to the output.

Definition 2: (Network) A network is the combination of a finite number of nodules. The output of the network is the sum of the outputs of all the nodules that constitute the network.

In practice all the networks of a population must have the same number of nodules, and this number, N, is fixed along the evolution.

Some parameters of the nodule are given by the problem and for that reason they are common to all the nodules: n, number of inputs; m, number of outputs; $x = (1, x_1, ..., x_n)$, input vector; f^{output}, transfer function of the output layer. The rest of the parameters depend on each nodule: h, number of (hidden) nodes of the nodule; f^i, transfer function of node i; p_i, partial output of node i (see explanation below); y_i, output of the node i; w_i, weight vector of node i.

As the node has a variable number of connections we have considered, for simplicity, that the connections that are not present in the node have weight 0, so we can use a weight vector of fixed length for all nodes. A node may have connections from input nodes, from other nodes and to output nodes. The weight vector is ordered as follows:

$$w_i = (\overbrace{w_{i,0}}^{bias}, \overbrace{w_{i,1}, \ldots, w_{i,n}}^{input}, \overbrace{w_{i,n+1}, \ldots, w_{i,n+h}}^{hidden}, \overbrace{w_{i,n+h+1}, \ldots, w_{i,n+h+m}}^{output}) \quad (1)$$

As there is no restriction in the connectivity of the nodule the transmission of the impulse along the connections must be defined in a way that avoids recurrence. The transmission has been defined in three steps:

Step 1. Each node generates its output as a function of only the inputs of the nodule (that is, the inputs of the whole network):

$$p_i = f^i\left(\sum_{j=0}^{n} w_{i,j} x_j\right), \quad (2)$$

this value is called partial output.

Step 2. These partial outputs are propagated along the connections. Then, each node generates its output as a function of all its inputs:

$$y_i = f^i\left(\sum_{j=0}^{n} w_{i,j} x_j + \sum_{j=1}^{h} w_{i,n+j} p_j\right). \quad (3)$$

Step 3. Finally, the output layer of the nodule generates its output:

$$o_j = f^{output}\left(\sum_{i=1}^{h} w_{i,n+h+j} y_i\right). \quad (4)$$

These three steps are repeated over all the nodules. The actual output vector of the network is the sum of the output vectors generated by each nodule.

Defined in this way a nodule is equivalent to a subnetwork of two hidden layers with the same number of nodes in both layers. This equivalent model is shown on Figure 9-2. So, the nodule of Figure 9-1 can be seen as the genotype of a nodule whose phenotype is the subnetwork shown on Figure 9-2. This difference is important, as the model of Figure 9-1 considered as a phenotype is a recurrent network.

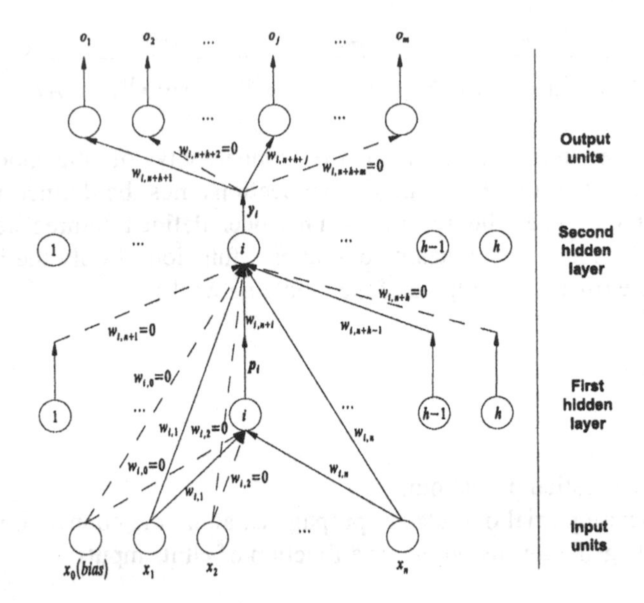

Figure 9-2. Equivalent model with two hidden layers. Every connection from an input node represents two connections, as the input value is used in two steps (see Equations 2 and 3). Every connection from another node of the nodule represents a connection between the first and second hidden layer (see Equation 3).

As the nodules must coevolve to develop different behaviours we have N_s independent subpopulations of nodules[10] that evolve separately. The network will always have N_s nodules, each one from a different subpopulation of nodules. Our task is not only to develop cooperative nodules but also to obtain the best combinations. For that reason we have also a population of networks. This population keeps track of the best combinations of nodules and evolves as the population of nodules evolves. The whole evolutionary process is shown in Figure 9-3.

Species creation is implicit, the subpopulations must coevolve complementary behaviours in order to get useful networks, as the combination of several nodules with the same behaviour when they receive the same inputs would not produce networks with a good fitness value. So, there is no need to introduce a mechanism for enforcing diversity that can bias the evolutionary process.

In the next two sections we will explain in depth the two populations and their evolutionary process.

[10] In order to maintain a coherent nomenclature we talk of one population of networks and another population of nodules. The population of nodules is divided into N_s genetically isolated subpopulations that coevolve together.

3.1.1 Nodule Population

The nodule population is formed by N_s subpopulations. Each subpopulation consists of a fixed number of nodules codified directly as subnetworks. The population is subject to the operations of replication and mutation. Crossover is not used due to its disadvantages in evolving artificial neural networks (Angeline et al., 1994). With these features the algorithm falls in the class of evolutionary programming (Fogel, 1992).

There is no limitation in the structure of the nodule or in the connections among the nodes. There is only one restriction to avoid unnecessary complexity in the resulting nodules, there can be no connections to an input node or from an output node.

The algorithm for the generation of a new nodule subpopulation is similar to other models proposed in the literature, such as GNARL (Angeline et al., 1994) or EPNet (Yao and Liu, 1997):

- The nodules of the initial subpopulation are created randomly. The number of nodes of the nodule, *h*, is obtained from a uniform distribution: $0 \leq h \leq h_{\max}$. Each node is created with a number of connections, *c*, taken from a uniform distribution: $0 \leq c \leq c_{\max}$. The initial value of the weights is uniformly distributed in the interval $[w_{\min}, w_{\max}]$.
- The new subpopulation is generated replicating the best *P%* of the former population. The remaining *(100-P)%* is removed and replaced by mutated copies of the best *P%*. An individual of the best *P%* is selected by roulette selection and mutated. This mutated copy substitutes one of the worst *(100-P)%* individuals.
- There are two types of mutation: parametric and structural. The severity of the mutation is determined by the relative fitness, F_r, of the nodule. Given a nodule υ its relative fitness is defined as:

$$F_r(\upsilon) = e^{-\alpha F(\upsilon)} \tag{5}$$

where $F(\upsilon)$ is the fitness value of nodule υ.

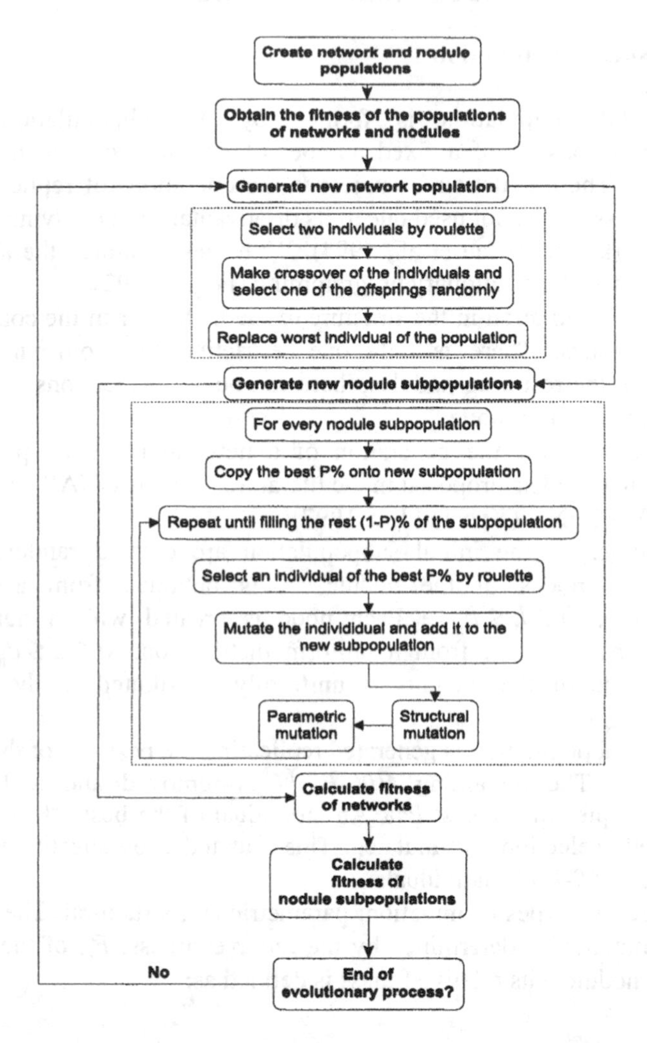

Figure 9-3. Evolutionary process of both populations. The generation of a new population for both populations, networks and nodules, is shown in detail.

Parametric mutation consists of a local search algorithm in the space of weights, a simulated annealing algorithm (Kirkpatrick et al., 1983). This algorithm performs random steps in the space of weights. Each random step affects all the weights of the nodule. For every weight, w_{ij}, of the nodule the following operation is carried out:

$$w_{ij} = w_{ij} + \Delta w_{ij}, \forall w_{ij} \in \upsilon, \tag{6}$$

where

$$\Delta w_{ij} \in N(0, \beta F_r(\upsilon)). \tag{7}$$

where β is a positive value that must be set by the user in order to avoid large steps in the space of weights. The value of β used in all our experiments has been $\beta=0.75$.

Then, the fitness of the nodule is recalculated and the usual simulated annealing criterion is applied. Being ΔF the difference in the fitness function before and after the random step:

- If $\Delta F \geq 0$ the step is accepted.
- If $\Delta F < 0$ then the step is accepted with a probability

$$P(\Delta F) = e^{-\frac{\Delta F}{T}},$$

where T is the current temperature. T starts at an initial value T_0 and is updated at every step, $T(t+1)=\gamma T(t)$, $0<\gamma<1$.

Parametric mutation is always carried out after structural mutation, as it does not modify the structure of the network.

Structural mutation is more complex because it implies a modification of the structure of the nodule. The behavioural link between parents and their offspring must be enforced to avoid generational gaps that produce inconsistency in the evolution. There are four different structural mutations: Addition of a node without connections, deletion of a node, addition of a connection with 0 weight, and deletion of a connection.

The nodes are added with no connections to enforce the behavioural link with its parent. As many authors have stated, (Yao and Liu, 1997; Angeline et al., 1994), maintaining the behavioural link between parents and their offspring is very important to get a useful algorithm.

All the above mutations are made in the mutation operation on the nodule. For each mutation there is a minimum value, Δ_m, and a maximum value, Δ_M. The number of elements (nodes or connections) involved in the mutation is calculated as follows:

$$\Delta = \Delta_m + F_r(\upsilon)(\Delta_M - \Delta_m). \tag{8}$$

So, before making a mutation the number of elements, Δ, is calculated, if $\Delta=0$ the mutation is not actually carried out.

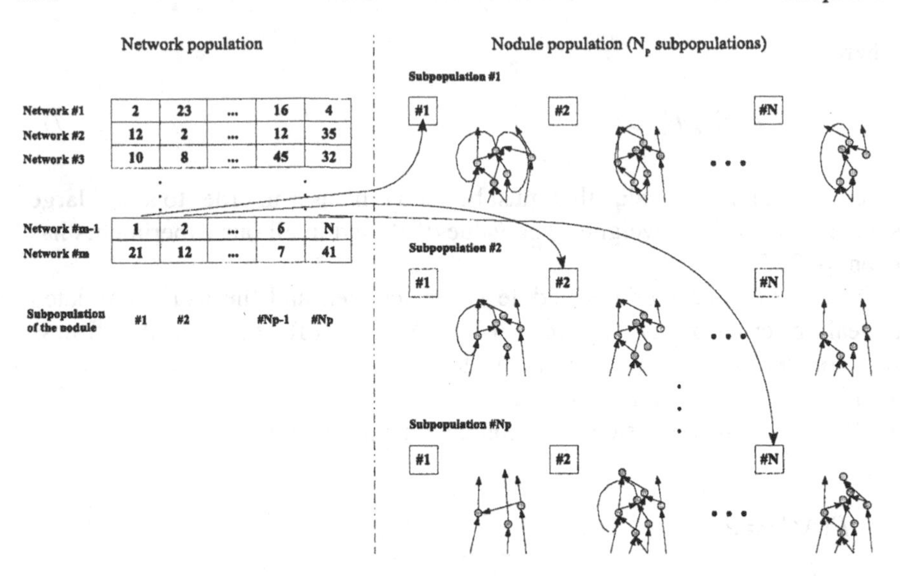

Figure 9-4. Populations of networks and nodules. Each element of the network is a reference to, or a label of, an individual of the corresponding subpopulation of nodules. So the network is a vector where the first component refers to a nodule of subpopulation 1, the second component to a nodule of subpopulation 2, and so on.

3.1.2 Network Population

The network population is formed by a fixed number of networks. Each network is the combination of one nodule of each subpopulation of nodules. So the networks are strings of integer numbers. The value of the numbers is not significant as they are just labels. The relationship between the two populations can be seen in Figure 9-4. It is important to note that, as the chromosome that represents the network is ordered, the permutation problem cannot appear.

The network population is evolved using the *steady-state* genetic algorithm (Whitley and Kauth, 1988; Whitley, 1989). This algorithm is selected because we need a population of networks that evolves more slowly than the population of nodules, as the changes in the population of networks have a major impact in the fitness of the nodules. The steady-state genetic algorithm avoids the negative effect that this drastic modification of the population of networks could have over the subpopulations of nodules. As the two populations evolve in synchronous generations, the modifications in the population of networks are less severe than the modifications in the subpopulations of modules.

In a steady-state genetic algorithm one member of the population is changed at a time. In the algorithm we have implemented the offspring generated by crossover replaces the two worst individuals of the population instead of replacing its parents. The algorithm allows adding mutation to the model.

Crossover is made at nodule level, using a standard two-point crossover. The parents exchange their nodules to generate their offspring. Mutation is also carried out at nodule level. When a network is mutated one of its nodules is substituted by another nodule of the same subpopulation.

During the generation of the new nodule population some nodules of every subpopulation are removed and substituted. The removed nodules are also substituted in the networks. This substitution has two advantages: first, poor performing nodules are removed from the networks and substituted by potentially better ones; second, the new nodules have the opportunity to participate in the networks immediately after their creation.

3.1.3 Fitness Assignment

The assignment of fitness to networks is straightforward. Each network is assigned a fitness in function of its performance in solving a given problem. If the model is applied to classification, the fitness of each network is the number of patterns of the training set that are correctly classified; if it is applied to regression, the fitness is the sum of squared errors, and so on.

Assigning fitness to the nodules is a much more complex problem. In fact, the assignment of fitness to the individuals that form a solution in cooperative evolution is one of its key topics. The performance of the model highly depends on that assignment. A discussion of the matter can be found in the Introduction of (Potter, 1997).

A credit assignment must fulfil the following requirements to be useful:

- It must enforce competition among the subpopulations to avoid subpopulations that develop similar responses to the same features of the data.
- It must enforce cooperation. The different subpopulations must develop complementary features that together can solve the problem.
- It must measure the contribution of a nodule to the fitness of the network, and not only the performance of the networks where the nodule is present. A nodule in a good network must not get a high fitness if its contribution to the performance of the network is not significant. Likewise, a nodule in a poor performing network must not be penalized if its contribution to the fitness of the network is positive. Otherwise, a good nodule that is temporarily assigned to poor rated networks may be lost in the evolution.

The method for calculating the fitness of the nodules consists of the weighted sum of three different criteria. These criteria, for obtaining the fitness of a nodule υ in a subpopulation π, are:

Substitution (σ) k networks are selected using an elitist method, that is, the best k networks of the population. In these networks the nodule of subpopulation π is substituted by the nodule υ. The fitness of the network with the nodule of the population π substituted by υ is measured. The fitness assigned to the nodule is the averaged difference in the fitness of the networks with the original nodule and with the nodule substituted by υ. This criterion enforces competition among nodules of the same subpopulation, as it tests if a nodule can achieve better performance than the rest of the nodules of its subpopulation.

The interdependencies among nodules could be a major drawback in the substitution criterion, but it does not mean that this criterion is useless. In any case, the criterion has two important features:

- It encourages the nodules to compete within the subpopulations, rewarding the nodules most compatible with the nodules of the rest of the subpopulation. This is true even for a distributed representation, because it has been shown that such representation is also modular. Moreover, as the nodules have no connection among them, they are more independent than in a standard network.
- As many of the nodules are competing with their parents, this criterion allows to measure if an offspring is able to improve the performance of its parents.

In addition, the neuropsychological evidence showing that certain parts of the brain consist of modules, that we discussed above, would support this objective.

Difference (δ) The nodule is removed from all the networks where it is present. The fitness is measured as the difference in performance of these networks. This criterion enforces competition among subpopulations of nodules preventing more than one subpopulation from developing the same behaviour. If two subpopulations evolve in the same way, the value of this criterion in the fitness of their nodules will be near 0.

Best k (β_k) The fitness is the mean of the fitness values of the best k networks where the nodule υ is present. Only the best k networks are selected because the importance of the worst networks of the population must not be significant. This criterion rewards the nodules in the best networks, and does not penalize a good nodule if it is in some poor performing network.

In order to encourage small nodules we have included a regularization term in the fitness of the nodule. Being n_n the number of nodes of the nodule and n_c the number of connections, the *effective fitness*[11], f_i', of the nodule is calculated following:

$$f_i' = f_i - \rho_n n_n - \rho_c n_c. \qquad (9)$$

The values of the coefficients must be in the interval $0<\rho_n, \rho_c<<1$ in order to avoid the regularization term introducing a high bias in the learning process.

So, the equation of the effective fitness of the nodule υ of subpopulation π is the following:

$$f_\upsilon^\pi = \lambda_\sigma \sigma + \lambda_\delta \delta + \lambda_{\beta_k} \beta_k - \rho_n n_n - \rho_c n_c, \qquad (10)$$

if the expression above is negative for any of the nodules of a subpopulation, then the fitness values of all the nodules of that subpopulation are shifted, as we have mentioned above, as following:

$$f_\upsilon^\pi = f_\upsilon^\pi - \min\{f_i^\pi\}_{i=1}^N, \qquad (11)$$

where N is the number of nodules of the nodule subpopulation.

Typical values of the weights of the components of the fitness used in our experiment are ($\lambda_\delta \approx 2\lambda_\sigma \approx 60\lambda_{\beta k}$). These values of the coefficients correct the differences in range of them.

4. EXPERIMENTS

The performance of the developed model is tested in ten classification problems with different features from the UCI Machine Learning Repository (Hettich et al., 1998). In order to get a clear idea of the performance of the model we have compared our model with a modular network, the *adaptive mixture of local experts* (Jacobs et al., 1991). Each *expert* is a multilayer perceptron (MLP) trained with standard back-propagation (Rumelhart et al., 1986) and a momentum term. We have also compared Covnet with the results in the literature.

For the design and training of the modular networks we have used *NeuralWorks Professional II/Plus* (NeuralWare, 1993) simulator. We also

[11] It is called effective fitness because it is the actual value used as the fitness of the nodule in the generation of a new subpopulation.

tried some pruning algorithms that are implemented in the Stuttgart Neural Network Simulator (SNNS)[12]: OBD (Le Cun et al., 1990), OBS (Hassibi and Stork, 1993) and Skeletonization (Mozer and Smolensky, 1989), but always with worse results.

Covnet has been programmed in C under the Linux Operating System. All the tools and programs used for its development are licensed under the GNU General Public License. Covnet's code[13] is also under the GNU General Public License.

All the parameters of Covnet are common to all the data sets used in the experiments. Such parameters are shown in Table 9-1. Setting the parameters for each problem improves the performance of Covnet but using the same parameters for all the problems shows the robustness of the model regarding the parameter setting.

Table 9-1. Covnet parameters common to all the experiments

Parameter	Value
Number of networks	100
Number of nodules on each subpopulation	40
Network to replace on each generation	2.0%
Mutation rate on network population	5.0%
Initial value of weights	(-0.5,0.5)
Nodule elitism	70%
Input scaling interval	[-2.5,2.5]
Number of nodule subpopulations	5
Initial maximum number of nodes	3
Initial maximum number of connections	15
Nodule fitness components	λ_σ=3.50
	λ_δ=1.45
	$\lambda_{\beta 3}$=0.05
Regularization term	ρ_n=0.25
	ρ_c=0.025
Simulated annealing	T_0=5.0
	α=0.95
	n=25
Minimum improvement (stop criterion)	10%

The parameters of the population, number of networks, number of nodule subpopulations and number of nodules per subpopulation, can have a variety of values. However, increasing the values shown in this chapter will not improve the performance, and will increase the computational cost of the evolution.

[12] This package could be obtained from ftp://ftp.informatik.uni-stuttgart.de/pub/SNNS by anonymous ftp.

[13] The code is available upon request to the authors.

The weight of the nodule fitness subcomponents must be fixed in a way that corrects the differences among their ranges. The values used in our experiments follow this idea. In a specific problem could be interesting considering any of the subcomponent more important than the others, but that can only be tested by trial and error.

Regularization parameters must be set in function of the importance of parsimony in our task. Increasing the values shown in this chapter will evolve smaller network, but also will decrease the performance of the networks as the regularization restriction becomes more critical.

At the end of the evolution the best network, in terms of training error, was selected as the result of the evolution. The test set was then used to obtain the generalization error of this network.

Each set of available data was divided into two sets: 75% of the patterns were used for learning and the remaining 25% for testing the generalization of the individuals. There are two exceptions, Sonar and Vowel problems, as the patterns of these two problems are prearranged in two subsets due to their specific features. Table 9-2 shows a summary of the data sets.

For training the modular networks we used the method of cross-validation and early-stopping (Finnoff et al., 1993). The networks were trained until the error over the validation set started to grow.

In all the tables we show, for each permutation of the data sets, the averaged error of classification over 30 runs of the algorithm, the standard deviation, the best and worst individuals, and the averaged number of nodes and connections of the best networks of each experiment. The measure of the error is the following:

$$E = \frac{1}{P}\sum_{i=1}^{P} e_i, \tag{12}$$

where P is the number of patterns and e_i is 0 if pattern i is correctly classified, and 1 otherwise.

The results obtained using Covnet and the modular neural network are shown in Table 9-3. In boldface we show the best results when the difference is statistically significant using a t-test at a confidence level of 95%. We can see that Covnet is able to outperform the modular network in 6 out of 10 datasets, and is worse only in 2 problems.

Table 9-2. Summary of data sets. The features of each data set can be C(continuous), B(binary) or N(nominal), nodes

Data set	Cases Train	Val	Test	Classes	Features C	B	N	Description
Cancer	360	175	174	2	9	-	-	There are two classes meaning if the cancer was benign (65.5% of the cases) or malignant (34.5%).
Card	346	172	172	2	6	4	5	There are two classes, meaning whether the application was granted (44.5% of the patterns) or denied (55.5%).
Gene	1588	794	793	3	-	-	60	This problem consists of two subtasks: recognizing exon/intron boundaries (referred to as EI sites), and recognizing intron/exon boundaries (IE sites).
Glass	107	54	53	6	9	-	-	This data set is from the UCI Machine Learning Repository. The set contains data from 6 different types of glass.
Heart	134	68	68	2	6	1	6	This data set comes from the Cleveland Clinic. The goal is the prediction of the presence or absence of heart disease in those patients.
Horse	182	91	91	3	13	2	5	This data set is from the UCI Machine Learning Repository. The aim is to predict the fate of a horse that has a colic: to survive, to die, of to be euthanized.
Pima	384	192	192	2	8	-	-	The patterns are divided into two classes that show whether the patient shows signs of diabetes.
Sonar	104	-	104	2	60	-	-	The task is to train a network to discriminate between sonar signals bounced off a metal cylinder and those bounced off a roughly cylindrical rock.
Soybean	342	171	170	19	6	13	16	The task is to recognize 19 different diseases of soy- beans.
Vowel	528	-	462	11	10	-	-	Speaker independent recognition of the eleven steady state vowels of British English.

The results obtained are good when they are compared with other works using these data sets. Table 9-4 shows a summary of the results reported in papers devoted to ensembles, modular networks or similar classification methods. Comparisons must be made cautiously, as the experimental setup is different in many papers. There are differences also in the methods used for estimating the generalization error. Some of the papers use 10-fold cross-validation that for some of the problems obtains a more optimistic estimation of the error.

Table 9-3. Error rates for the modular network and Covnet for the datasets. The best generalization error is in boldface for every problem

Problem	Modular network			Covnet		
	Mean	Best	Worst	Mean	Best	Worst
Cancer	0.0224	0.0115	0.0345	0.0167	0.0105	0.0230
Card	0.1374	0.1163	0.1686	**0.1157**	0.0930	0.1395
Gene	0.1511	0.1324	0.1702	**0.1398**	0.1021	0.1425
Glass	0.3904	0.0745	0.1180	**0.3723**	0.1321	0.3019
Heart	0.1941	0.1324	0.2500	**0.1426**	0.0882	0.2059
Horse	**0.2714**	0.2308	0.3077	0.2780	0.2308	0.3407
Pima	0.2299	0.1771	0.2865	**0.1990**	0.1615	0.2448
Sonar	**0.1875**	0.1346	0.2506	0.2202	0.2019	0.2404
Soybean	0.2023	0.1235	0.2765	**0.1985**	0.1235	0.2412
Vowel	0.5821	0.5325	0.6710	0.5788	0.4913	0.5190

Table 9-4. Results of previous works using the same data sets. We record the results of the best method among the algorithms tested in each paper

Data set	Reference												
	Coop	(Yao and Liu, 1998)[1]	(Webb, 2000)[2]	(Zenobi and Cunningham, 2001)[3]	(Merz, 1999)	(Friedman et al., 2000)[2]	(Liu et al., 2000)[2]	(Liu et al., 2001)[2]	(Islam et al., 2003)[1]	(Dietterich, 2000)[2]	(Dzeroski and Zenko, 2004)[2]	(Breiman, 2000)[1]	(Todorovski and Dzeroski, 2003)[2]	(Cantú-Paz and Kamath, 2003)[1]
Cancer	0.0123	-	0.035	-	-	0.038	-	-	0.0120	0.0310	0.0272	0.034	0.0263	0.033
Card	0.1217	0.093	-	-	0.1398	-	0.135	0.130	0.0910	0.130	0.1432	-	0.1433	-
Gene	0.1238	-	-	-	-	-	-	-	-	0.0503	-	0.051	-	-
Glass	0.2289	-	0.249	-	0.3144	0.238	-	-	0.2518	0.2277	0.2519	0.226	0.3154	0.329
Heart	0.1196	0.151	0.197	0.166	0.1751	-	-	-	0.1384	0.2045	0.1604	-	0.1604	-
Horse	0.2674	-	0.169	-	-	-	-	-	-	0.1825	-	-	-	-
Pima	0.1969	0.226	0.244	0.234	-	-	0.221	0.223	0.1960	-	0.2402	-	0.2372	0.260
Sonar	0.1436	-	-	-	0.2278	0.154	-	-	-	0.1651	0.1529	0.163	-	-
Soybean	0.0761	-	0.070	-	-	-	-	-	0.0781	0.0757	0.0633	0.056	0.0568	-
Vowel	0.4587	-	-	-	-	0.517[1]	-	-	-	-	-	-	-	-

[1] Hold out.

[2] k-fold cross-validation.

[3] Best classifier.

5. CONCLUSIONS

In this chapter we have shown how a cooperative coevolutionary model for the design of artificial neural networks can be developed. This model is based on the coevolution of several species of subnetworks that must cooperate to form networks for solving a given problem. Instead of trying to evolve whole networks, a task that is not feasible in many problems or ends up with poorly performing neural networks, we evolve these subnetworks that must cooperate in solving the given task. The nodules coevolve in several independent subpopulations that evolve to different species. A population of networks that is evolved by means of a steady-state genetic algorithm keeps track of the best combinations of nodules for solving the problem.

We have also presented a new method for assigning credit to the individuals of the different species that cooperate to form a network. This method is based on the combination of three criteria. The criteria enforce competition within species and cooperation among species. The same idea underlying this method could be applied to other models of cooperative coevolution.

This model has proved to perform better than standard algorithms in ten real problems of classification. Moreover, it has shown better results than the methods of training modular neural networks by means of gradient descent, e.g. the backpropagation learning rule, in 6 out of 10 problems.

Networks evolved by Covnet are very compact and have few sparsely distributed connections. These networks are appropriate for hardware implementation.

ACKNOWLEDGMENTS

The authors would like to acknowledge R. Moya-Sánchez for her helping in the final version of this chapter. Part of the work reported in this chapter has been financed by the Project TIC2002-04036-C05-02 of the Spanish CICYT and FEDER funds.

REFERENCES

Angeline, P. J., Saunders, G. M., and Pollack, J. B., 1994, An evolutionary algorithm that constructs recurrent neural networks, *IEEE Transactions on Neural Networks* **5**(1):54-65.

Belew, R. K., McInerney, J., and Schraudolph, N. N., 1991, Evolving networks: Using genetic algorithms with connectionist learning, Tech. Rep. CS90-174, Computer Science Engineering Department, University of California-San Diego.

Borst, M. V., 1994, *Local Structure Optimization in Evolutionary Generated Neural Network Architectures*, Ph.D. Thesis, Leiden University, The Netherlands.

Breiman, L., 2000, Randomizing outputs to increase prediction accuracy, *Machine Learning* **40**:229-242.

Caelli, T., Guan, L., and Wen, W., 1999, Modularity in neural computing, *Proceedings of the IEEE* **87**(9):1497-1518.

Cantú-Paz, E., and Kamath, C., 2003, Inducing oblique decision trees with evolutionary algorithms, *IEEE Transactions on Evolutionary Computation* **7**(1):54-68.

Chellapilla, K., and Fogel, D. B., 1999, Evolving neural networks to play checkers without relying on expert knowledge, *IEEE Transactions on Neural Networks* **10**(6):1382-1391.

Cho, S-B., and Shimohara, K., 1998, Evolutionary learning of modular neural networks with genetic programming, *Applied Intelligence* **9**:191-200.

Depenau, J., and Moller, M., 1994, Aspects of generalization and pruning, in: *Proc. World Congress on Neural Networks*, vol. III, pp. 504-509.

Dietterich, T. G., 2000, An experimental comparison of three methods for constructing ensembles of decision trees: Bagging, boosting, and randomization, *Machine Learning* **40**:139-157.

Dzeroski, S., and Zenko, B., 2004, Is combining classifiers with stacking better than selecting the best one?, *Machine Learning* **54**:255-273.

Finnoff, W., Hergert, F., and Zimmermann, H. G., 1993, Improving model selection by nonconvergent methods, *Neural Networks* **6**:771-783.

Fogel, D. B., 1992, *Evolving Artificial Intelligence*, Ph.D. thesis, University of California, San Diego.

Friedman, J., Hastie, T., and Tibshirani, R., 2000, Additive logistic regression: A statistical view of boosting, *Annals of Statistics* **28**(2), pp. 337-407.

Gallant, S., 1993, *Neural-Network Learning and Expert Systems*, MIT Press, Cambridge, MA.

García-Pedrajas, N., Hervás-Martínez, C., and Muñoz-Pérez, J., 2003, Covnet: A cooperative coevolutionary model for evolving artificial neural networks, *IEEE Transactions on Neural Networks* **14**(3):575-596.

Goldberg, D. E., 1989a, Genetic algorithms and Walsh functions: Part 1, a gentle introduction, *Complex Systems* **3**:129-152.

Goldberg, D. E., 1989b, Genetic algorithms and Walsh functions: Part 2, deception and its analysis, *Complex Systems* **3**:153-171.

Goldberg, D. E., 1989c, Genetic Algorithms in Search, *Optimization and Machine Learning*, Addison-Wesley, Reading, MA.

Goldberg, D., and Deb, K., 1991, A comparative analysis of selection schemes used in genetic algorithms, in: *Foundations of Genetic Algorithms*, G. Rawlins, ed., Morgan Kaufmann, pp. 94-101.

Hancock, P. J. B., 1992, Genetic algorithms and permutation problems: A comparison of recombination operators for neural net structure specification, in: *Proc. Int. Workshop on Combinations of Genetic Algorithms and Neural Networks* (COGANN-92), D. Whitley and J. D. Schaffer, eds., Los Alamitos, CA, IEEE Computer Soc. Press, pp. 108-122.

Hassibi, B., and Stork, D., 1993, Second order derivatives for network pruning: Optimal brain surgeon, in *Advances in Neural Information Systems* **5**:164-171.

Haykin, S., 1994, *Neural Networks—A Comprehensive Foundation*, Macmillan College Publishing Company, New York, NY.

Hettich, S., Blake, C. L., and Merz, C. J., 1998, UCI repository of machine learning databases, http://www.ics.uci.edu/~mlearn/MLRepository.html.

Hirose, Y., Yamashita, K., and Hijiya, S., 1991, Backpropagation algorithm which varies the number of hidden units, *Neural Networks* **4**:61-66.

Honavar, V., and Uhr, V. L., 1993, Generative learning structures for generalized connectionist networks, *Information Science* **70**(1):75-108.

Islam, Md. M., Yao, X., and Murase, K., 2003, A constructive algorithm for training cooperative neural network ensembles, *IEEE Transactions on Neural Networks* **14**(4):820-834.

Jacobs, R. A., Jordan, M. I., Nowlan, S. J., and Hinton, G. E., 1991, Adaptive mixtures of local experts, *Neural Computation* **3**:79-87.

Kamimura, R., and Nakanishi, S., 1994, Weight-decay as a process of redundancy reduction, in: *Proceedings of World Congress on Neural Networks*, vol. III, pp. 486-489.

Kirkpatrick, S., Gelatt Jr, C. D., and Vecchi, M. P., 1983, Optimization by simulated annealing, *Science* **220**:671-680.

Le Cun, Y., Denker, J. S., and Solla, S. A., 1990, Optimal brain damage, in: *Advances in Neural Information Processing* (2), D. S. Touretzky, ed., Denver, CO, pp. 598-605.

Lin, Ch-T., and Jou, Ch-P., 1999, Controlling chaos by GA-based reinforcement learning neural network, *IEEE Transactions on Neural Networks* **10**(4):846-859.

Liu, Y., and Yao, X., 1999, Ensemble learning via negative correlation, *Neural Networks* **12**(10):1399-1404.

Liu, Y., Yao, X., and Higuchi, T., 2000, Evolutionary ensembles with negative correlation learning, *IEEE Transactions on Evolutionary Computation* **4**(4):380-387.

Liu, Y., Yao, X., Zhao, Q., and Higuchi, T., 2001, Evolving a cooperative population of neural networks by minimizing mutual information, in: *Proc. of the 2001 IEEE Congress on Evolutionary Computation*, Seoul, Korea, pp. 384-389.

Maniezzo, V., 1994, Genetic evolution of the topology and weight distribution of neural networks, *IEEE Transactions on Neural Networks* **5**(1):39-53.

Merz, C. J., 1999, Using correspondence analysis to combine classifiers, *Machine Learning* **36**(1):33-58.

Michalewicz, Z., 1994, *Genetic Algorithms + Data Structures = Evolution Programs*, Springer-Verlag, New York.

Miller, G. F., Todd, P. M., and Hedge, S. U., 1991, Designing neural networks, *Neural Networks* **4**:53-60.

Moriarty, D. E., 1997, *Symbiotic Evolution of Neural Networks in Sequential Decision Tasks*, Ph.D. thesis, University of Texas at Austin, Tech. Rep. AI97-257.

Moriarty, D. E., and Miikkulainen, R., 1996, Efficient reinforcement learning through symbiotic evolution, *Machine Learning* **22**:11-32.

Mozer, M. C., and Smolensky, P., 1989, Skeletonization: A technique for trimming the fat from a network via relevance assessment, in: *Advances in Neural Information Processing* (1), D. S. Touretzky, Ed., Denver, CO, pp. 107-155.

NeuralWare, 1993, Neural Computing: A Technology Handbook for Professional II/Plus, NeuralWare Inc., Pittsburgh, PA.

Odri, S. V., Petrovacki, D. P., and Krstonosic, G. A., 1993, Evolutional development of a multilevel neural network, *Neural Networks* **6**:583-595.

Opitz, D. W., and Shavlik, J. W., 1996, Actively searching for an effective neural network ensemble, *Connection Science* **8**(3):337-353.

Parekh, R., Yang, J., and Honavar, V., 2000, Constructive neural-network learning algorithms for pattern classification, *IEEE Transactions on Neural Networks* **11**(2):436-450.

Potter, M. A., 1997, *The Design and Analysis of a Computational Model of Cooperative Coevolution*, Ph.D. Thesis, Goerge Mason University, Fairfax, Virginia.

Potter, M. A., and de Jong, K. A., 2000, Cooperative coevolution: An architecture for evolving coadapted subcomponents, *Evolutionary Computation* **8**(1):1-29.

Reed, R., 1993, Pruning algorithms - A survey, *IEEE Transactions on Neural Networks* **4**:740-747.

Rosen, B. E., 1996, Ensemble learning using decorrelated neural networks, *Connection Science* **8**(3):373-384.

Rumelhart, D., Hinton, G., and Williams, R. J., 1986, Learning internal representations by error propagation, in: *Parallel Distributed Processing*, D. Rumelhart and J. McClelland, eds., MIT Press, Cambridge, MA, pp. 318-362.

Samuel, A. L., 1959, Some studies in machine learning using the game of checkers, *Journal of Research and Development* **3**(3):210-229.

Schaffer, J. D., Whitley, L. D., and Eshelman, L. J., 1992, Combinations of genetic algorithms and neural networks: A survey of the state of the art, in: *Proc. Int. Workshop on Combinations of Genetic Algorithms and Neural Networks* (COGANN-92), D. Whitley and J. D. Schaffer, eds., Los Alamitos, CA, pp. 1-37, IEEE Computer Soc. Press.

Shang, Y., and Wah, B. W., 1996, Global optimization for neural networks training, *IEEE Computer* **29**(3):45-54.

Smalz, R., and Conrad, M., 1994, Combining evolution with credit apportionment: A new learning algorithm for neural nets, *Neural Networks* **7**(2):341-351.

Thodberg, H. H., 1991, Improving generalization of neural networks through pruning, *International Journal of Neural Systems* **1**(4):317-326.

Todorovski, L., and Dzeroski, S., 2003, Combining classifiers with meta decision trees, *Machine Learning* **50**:223-249.

van Rooij, A. J. F., Jain, L. C., and Johnson, R. P., 1996, *Neural Networks Training Using Genetic Algorithms*, vol. 26 of Series in Machine Perception and Artificial Intelligence, World Scientific, Singapore.

Webb, G. I., 2000, Multiboosting: A technique for combining boosting and wagging, *Machine Learning* **40**(2):159-196.

Whitehead, B. A., and Choate, T. D., 1996, Cooperative-competitive genetic evolution of radial basis function centres and widths for time series prediction, *IEEE Transactions on Neural Networks* **7**(4):869-880.

Whitley, D., 1989, The GENITOR algorithm and selective pressure, in: *Proc 3rd International Conf. on Genetic Algorithms*, Morgan Kaufmann Publishers, Los Altos, CA, pp. 116-121.

Whitley, D., and Kauth, J., 1988, GENITOR: A different genetic algorithm, in: *Proceedings of the Rocky Mountain Conference on Artificial Intelligence*, Denver, CO, pp. 118-130.

Yao, X., 1999, Evolving artificial neural networks, *Proceedings of the IEEE* **9**(87):1423-1447.

Yao, X., and Liu, Y., 1997, A new evolutionary system for evolving artificial neural networks, *IEEE Transactions on Neural Networks* **8**(3):694-713.

Yao, X., and Liu, Y., 1998, Making use of population information in evolutionary artificial neural networks, *IEEE Transactions on Systems, Man, and Cybernetics—Part B: Cybernetics* **28**(3):417-425.

Zenobi, G., and Cunningham, P., 2001, Using diversity in preparing ensembles of classifiers based on different feature subsets to minimize generalization error, in: *12th European Conference on Machine Learning* (ECML 2001), L. de Raedt and P. Flach, eds., Lecture Notes in Artificial Intelligence, vol. 2167, Springer-Verlag, pp. 576-587.
Zhao, Q. F., Hammami, O., Kuroda, K., and Saito, K., 2000, Cooperative co-evolutionary algorithm—How to evaluate a module?, in: *Proc. 1st IEEE Symposium of Evolutionary Computation and Neural Networks*, San Antonio, TX, pp. 150-157.

Chapter 10

GREEDY RANDOMIZED ADAPTIVE SEARCH PROCEDURES

Francisco R. Angel-Bello[1], José Luis González-Velarde[1] and Ada M. Alvarez[2]

[1]*Instituto Tecnológico de Estudios Superiores de Monterrey, Nuevo León, México;* [2]*Universidad Autónoma de Nuevo León, Nuevo León, México*

Abstract: This chapter addresses the problem of designing and training artificial neural networks with discrete activation functions for the classification of patterns in two categories. When the set of patterns is not linearly separable, this problem consists of determining the amount of neurons in the hidden layer that are needed to correctly classify the patterns. This problem has been reported to be an *NP-hard* problem. In this case, a GRASP is proposed, which exploits the particular structure of the model to determine the neurons of the hidden layer of the network as well as their corresponding weights. This procedure adds neurons, one at a time, until there are no misclassified patterns. Then it is possible to apply the condition of linear separability in order to obtain the corresponding weights of the neuron of the output layer. As a result, a trained network is obtained, which correctly classifies all the patterns in the training set. The procedure is tested with ten benchmarks datasets and results show that it performs well in a reasonable amount of time.

Key words: Classification problem; neural networks; GRASP; constructive procedure.

1. INTRODUCTION

The goal of this chapter is to show how GRASP can be used to design a neural network for the classification or recognition problem, which consists of identifying the class to which a given object belongs.

Designing/training feedforward neural networks involves the determination of the number of neurons in the hidden layer. The universal approximation theorem (Cybenko, 1989; Hornik, 1989) establishes that a single hidden layer is sufficient for a multilayer perceptron to compute a uniform approximation to a given training set represented by a set of inputs and a desired output. This is known as the universal approximator property. This theorem is important from a theoretical viewpoint, but is not constructive, that is, it does not specify how to determine the minimum number of neurons in the hidden layer to reach the approximation.

There are related papers that report upper bounds for the number of neurons in the hidden layer so that the learning algorithm (for example, backpropagation) is convergent (Gori, 1992; Yu, 1992). Nevertheless, in general, these values are too large and it has been verified in several applications that the algorithm converges with a smaller number of neurons. Moreover, a neural network with minimum size is less likely to learn the noise in the training data, and may thus generalize better to new data.

The most popular methods, used to determine the number of neurons in the hidden layer, are the pruning and growth procedures. Pruning procedures start with a large multiplayer perceptron with an adequate performance for the problem at hand, and then prune it by weakening or eliminating certain synaptic weights in a selective and orderly fashion. In the other hand, growing procedures start with a small multiplayer perceptron for accomplishing the task at hand, and then add the neuron or a new layer of hidden neurons only when it is not possible to meet the design specifications. Both procedures work in a trial-and-error way, so the process of training is computationally expensive.

Another way to approach the problem is by using constructive learning algorithms (Parekh et al., 2000). These learning algorithms offer an attractive framework for the incremental construction of near-minimal neural network architectures in the sense of the number of neurons in the hidden layers. Their operations are based on different modifications of the learning algorithm for the discrete perceptron and the goal is to determine, in each iteration of the algorithm, the larger linearly separable subset from a given training set.

Fletcher and Obradovic (1994) proposed another constructive algorithm, which determines separating hyperplanes by repeatedly interpolating between example points of both classes. However, this algorithm requires significant computational resources.

Angel-Bello et al. (2003) presented a linear-programming-based heuristic for designing and training neural networks with discrete activation function for pattern recognition. This procedure exploits the particular structure of the model to determine the neurons of the hidden layer of the network as well as their corresponding weights.

In the present work a greedy randomized adaptive algorithm is proposed which uses linear formulations to determine, one at a time, hyperplanes (neurons) that separate the largest possible amount of patterns of a same class. Consequently, the number of training patterns decreases successively, until a linearly separable subset is obtained.

The chapter is organized as follows: Section 2 describes the metaheuristic GRASP generalities, the mathematical formulation and some theoretical aspects are established in Section 3, in Section 4 the proposed implementation of GRASP to the problem addressed is presented, Section 5 presents the computational experiment and Section 6 the conclusions and future works.

2. GRASP GENERALITIES

A greedy randomized adaptive procedure (GRASP) is a metaheuristic that has been increasingly used in diverse applications because of its simplicity and effectiveness.

GRASP is a multi-start procedure, which consists basically of two phases: construction and post-processing [Feo and Resende 1995]. The construction phase tries to build a solution step by step, adding one element at a time, while the post-processing phase attempts to improve this solution. At the beginnings, only local search was tried, but nowadays other techniques as Scatter Search have been increasingly used to improve the solutions. The process is repeated many times and the best overall solution is kept as the result.

The construction phase is characterized by three main features:

(a) A greedy function: It is used to determine which candidate element should be added to the partial solution. This function measures the benefit of selecting each element to be included in the solution, that is, measures the local contribution of each candidate element to the partial solution.

(b) Randomized selection: All the elements that are candidate to be included in the solution are ordered in a list with respect to the values of the greedy function. Then, one element is selected among the best candidates in the list, but not necessarily the top candidate. This choice technique allows the procedure to be repeated in order to generate different solutions. The list of best candidates is called *restricted candidate list* (RCL) and may be formed by a fixed number of candidates (cardinality restriction) or by those elements whose values of the greedy function lie in a pre-specified range (value restriction).

(c) Adaptive feature: The procedure is adaptive because the benefit (greedy function value) of each element is updated at each iteration of the construction phase in order to incorporate the new information that is gathered after the selection of the previous element.

The solutions generated by a GRASP construction are not guaranteed to be locally optimal with respect to simple neighborhood definitions. Hence, it is almost always recommended to apply a local search or other technique to attempt to improve each constructed solution.

It is easy to implement a GRASP. There are few parameters that need to be set and tuned (number of constructions, length of RCL) and therefore one can focus on implementing efficient data structures to assure quick GRASP iterations. Besides, GRASP can be trivially implemented in parallel. Each processor can be initialized with its own copy of the procedure, the instance data and an independent random number sequence. The GRASP iterations are then performed in parallel and only a single global variable is required to store the best solution found overall processors.

Although GRASP has been used in a wide range of applications (Festa and Resende, 2002), to the best of the authors knowledge, this is the first time it has been used to design and train neural networks.

3. MATHEMATICAL FOUNDATIONS

The problem of classifying patterns into two classes can be stated as follows:

Given a pattern set $X=\{X_1,X_2,\cdots,X_T\}$, there is a value $d_i \in \{-1, 1\}$ associated to each pattern $X_i=(x_{i1},x_{i2},\cdots,x_{in})^t \in \mathbf{R}^n$. This value represents the pattern's membership to any of the classes, according to the following rule:

$$d_i = \begin{cases} 1, & \text{if } X_i \in C_1 \\ -1, & \text{if } X_i \in C_2 \end{cases}$$

DEFINITION:

The set of patterns $X=\{X_1,X_2,\cdots,X_T\}$ is linearly separable if there exists a vector $W=(w_1,w_2,\cdots,w_{n+1})^t \in \mathbf{R}^{n+1}$, such that

$$\begin{cases} \sum_{j=1}^{n} x_{ij} w_j - w_{n+1} > 0, \text{for } X_i \in C_1 \\ \sum_{j=1}^{n} x_{ij} w_j - w_{n+1} < 0, \text{for } X_i \in C_2 \end{cases} \tag{1}$$

Multiplying each inequality by its respective value of d_i, the Eq. (1) is transformed to

$$d_i \left[\sum_{j=1}^{n} x_{ij} w_j - w_{n+1} \right] > 0 \tag{2}$$

On the other hand, if there exists a vector of weights $W \in \mathbf{R}^{n+1}$ such that Eq. (2) holds, then it is possible, by an appropriate scaling, to find a vector $\hat{W} \in \mathbf{R}^{n+1}$ so as to make the following expression true.

$$d_i \left[\sum_{j=1}^{n} x_{ij} \hat{w}_j - \hat{w}_{n+1} \right] \geq 1 \tag{3}$$

Consequently, the problem of investigating whether a set of patterns is linearly separable has been transformed to the problem of determining if there exists feasible solution to the system represented by Eq. (3).

From (3) the following result can be stated, which represents a necessary and sufficient condition for determining whether a set of patterns is linearly separable.

PROPOSITION:

The set of patterns $X = \{X_1, X_2, \cdots, X_T\}$ is linearly separable by the hyperplane $\sum_{j=1}^{n} x_{ij} w_j - w_{n+1} = 0$, if and only if the objective value of the following linear program is zero.

$$\min z = \sum_{i=1}^{T} z_i$$

$$s.t.\begin{cases} d_i\left[\sum_{j=1}^{n} x_{ij} w_j - w_{n+1}\right] + z_i \geq 1, (i = 1,2,\cdots,T) \\ z_i \geq 0, (i = 1,2,\cdots,T) \end{cases} \tag{P1}$$

It is clear that the objective function only takes value zero when all the variables z_i are equal to zero and in this case there exist vectors W such that Eq. (3) holds. Thus, the set of patterns $X = \{X_1, X_2, \cdots, X_T\}$ is linearly separable.

On the other hand, if the set of patterns $X = \{X_1, X_2, \cdots, X_T\}$ is linearly separable, then Eq. (3) is satisfied, and it is possible to conclude that the variables z_i are not necessary to achieve feasibility in the linear programming problem's constraints. Thus, the optimum is achieved when all these variables are equal to zero and in this case the objective value $z = 0$.

Using this proposition it can be verified, in an efficient way, if a set of patterns is separable or not. In the case that the set is not separable, the system represented by Eq. (3) is not feasible and any feasible solution to problem (P1) contains variables z_i with positive values, corresponding to

that patterns that are misclassified by hyperplane $\sum_{j=1}^{n} x_{ij} w_j - w_{n+1} = 0$.

Taking into account that variables z_i in (P1) are equivalent to the artificial variables in the phase 1 of the Simplex method, it may occur that the minimum is reached with all these variables as basic variables. This solution, feasible to (P1), does not have any meaning when we are determining a separating hyperplane.

To override this and to determine the number of patterns that are

misclassified by the hyperplane $\sum_{j=1}^{n} x_{ij} w_j - w_{n+1} = 0$, the following problem

can be formulated:

$$
\min z = \sum_{i=1}^{T} y_i
$$

$$
s.t.\begin{cases} d_i\left[\sum_{j=1}^{n} x_{ij} w_j - w_{n+1}\right] + z_i \geq 1 \\ z_i \leq M y_i \\ z_i \geq 0,\ y_i \in \{0,1\} \\ (i = 1,2,\cdots,T) \end{cases} \tag{IP1}
$$

where M is a very large constant and $y_i = \begin{cases} 1, & \text{si } z_i > 0 \\ 0, & \text{si } z_i = 0 \end{cases}$

It has been proved (Roychowdhury et al., 1995) that when the set of patterns is not linearly separable, the problem of determining the maximum cardinality subset, which is linearly separable, is an ***NP-complete*** problem. Therefore, in order to find separating hyperplanes, which solve the classification problem, a heuristic algorithm will be designed.

4. SOLUTION APPROACH

Solving problem (IP1) we obtain a hyperplane $\sum_{j=1}^{n} x_{ij} w_j - w_{n+1} = 0$ that correctly classifies the maximum number of patterns. This hyperplane partitions the pattern set in two subsets that may contain patterns belonging to both classes. Then, for each subset one would formulate problems similar to (IP1) and obtain two hyperplanes that partition these subsets. One may continue in the same way until obtain linearly separable subsets and as a consequence the neurons of the hidden layer. The generated hyperplanes are the hidden neurons.

The disadvantage of using this approach to design the network is that it is always necessary to consider the whole pattern set even when they are grouped in smaller subsets. Note that we are considering the pattern set as the training set.

To override this, we will impose the constraint that all the patterns in one of the half-space defined by the hyperplane belong to the same class. Consequently, the size of the problem to be solved decreases from one iteration to the next. This would be the adaptive feature of the procedure.

The problems to determine a hyperplane that separates the maximum number of patterns belonging to one class can be formulated as follows:

$$\min z = \sum_{i=1}^{T_1} y_i$$

$$s.t.\begin{cases} d_i\left[\sum_{j=1}^{n} x_{ij}w_j - w_{n+1}\right] + z_i \geq 1, (i = 1,2,\cdots,T_1) \\ d_i\left[\sum_{j=1}^{n} x_{ij}w_j - w_{n+1}\right] \geq 1, (i = T_1+1,\cdots,T) \\ z_i \leq My_i, (i = 1,2,\cdots,T_1) \\ z_i \geq 0,\ y_i \in \{0,1\}, (i = 1,2,\cdots,T_1) \end{cases} \tag{IP2}$$

$$\min z = \sum_{i=T_1+1}^{T} y_i$$

$$s.t.\begin{cases} d_i\left[\sum_{j=1}^{n} x_{ij}w_j - w_{n+1}\right] \geq 1, (i = 1,2,\cdots,T_1) \\ d_i\left[\sum_{j=1}^{n} x_{ij}w_j - w_{n+1}\right] + z_i \geq 1, (i = T_1+1,\cdots,T) \\ z_i \leq My_i, (i = T_1+1,\cdots,T) \\ z_i \geq 0,\ y_i \in \{0,1\}, (i = T_1+1,\cdots,T) \end{cases} \tag{IP3}$$

Solving these problems we obtain hyperplanes that separate the maximum number of patterns belonging to one class, but empirical evidence indicates that choosing the best solution, that is, the hyperplane that separates the maximum number of patterns do not guarantee to obtain a neural network with the minimum number of neurons in the hidden layer. For this reason we will look for different feasible solutions to these problems.

To find feasible solutions to problems (IP2) and (IP3), the following problems are formulated:

$$\min z = \sum_{i=1}^{T_1} z_i$$

$$s.t. \begin{cases} d_i \left[\sum_{j=1}^{n} x_{ij} w_j - w_{n+1} \right] + z_i \geq 1, (i = 1, 2, \cdots, T_1) \\ d_i \left[\sum_{j=1}^{n} x_{ij} w_j - w_{n+1} \right] \geq 1, (i = T_1 + 1, \cdots, T) \\ z_i \geq 0, (i = 1, 2, \cdots, T_1) \end{cases} \tag{P2}$$

$$\min z = \sum_{i=T_1+1}^{T} z_i$$

$$s.t. \begin{cases} d_i \left[\sum_{j=1}^{n} x_{ij} w_j - w_{n+1} \right] \geq 1, (i = 1, 2, \cdots, T_1) \\ d_i \left[\sum_{j=1}^{n} x_{ij} w_j - w_{n+1} \right] + z_i \geq 1, (i = T_1 + 1, \cdots, T) \\ z_i \geq 0, (i = T_1 + 1, 2, \cdots, T) \end{cases} \tag{P3}$$

In order to obtain different solutions, one pattern will be enforced to be well classified, fixing its associated $z_i = 0$. In the case that feasible solution exists, it gives a hyperplane that separates certain number of patterns. Note that for all of these patterns do not make sense to fix in zero its associated z_i because we would obtain the same solution. Then, we are going to fix in zero those variables z_i that correspond to patterns that have not been before separated.

In this way, hyperplanes that separate different number of patterns are obtained and it is possible to sort them with respect to this number and randomly choose one of the best evaluated.

It has been empirically shown that choosing the best solution do not guarantee that a neuronal network with the less number of neurons in the hidden layer will be obtained. So, it seems reasonable to incorporate the foregoing derived results in a procedure, such as GRASP, that can take advantage from this fact.

4.1 Description of the GRASP to Design/Train the Neural Network

As mentioned before, each GRASP iteration consists basically of two phases: construction and post-processing. In this implementation post-processing is not performed, and GRASP is restricted to the construction phase, which tries to build a solution (a neural network) step by step, adding one element (one hyperplane/neuron in this case) at a time. As have been said before, this phase is characterized by a greedy adaptive measure and a randomized selection.

The greedy measure that will evaluate the benefit of incorporating a determined selection, in our case, a hyperplane, in a solution being created is the number of patterns that the hyperplane separates.

In order to choose a hyperplane/neuron, a set of possible neurons should be previously found. From this set, a restricted candidate list will be drawn and a fixed length of RCL will be considered in this work. The selection of a neuron to be included in a network is randomly made from this candidate list. This mechanism allows the procedure to be repeated in order to generate different solutions.

Once a hyperplane has been added to the partially constructed solution, the values of the greedy function must be reevaluated. This makes the procedure acquire the *adaptive* feature that characterizes it.

Each GRASP construction can be sketched as follows. The input is a set of patterns:

1. Formulate problems (P2) and (P3) and randomly choose one of them to be solved.

2. Solve the selected problem fixing all variables $z_i = 0$, one at a time. The solutions of this problem indicate the values of the weights and the threshold of each hyperplane and the number of patterns that they separate.

3. Build the RCL with the set of hyperplanes obtained in step 2.

4. Select randomly one hyperplane from RCL and add it to the network.

5. Eliminate all the patterns that were separated by the added hyperplane. As long as there are patterns from both classes, formulate accordingly problems (P2) and (P3) and go to step 1, otherwise stop.

Note that besides the random selection of the hyperplane from RCL, there is an extra random element in the procedure: the selection of the problem to be solved. Solving any of these linear problems we obtain a separating hyperplane which the following characteristic: in one of the half-spaces there are only patterns belonging to the same class, while in the other half-space we can find patterns belonging to both classes. When problem (P2) is solved the obtained hyperplane separates patterns from class 1, while when (P3) is solved the hyperplane separates patterns from class 2.

The matrix $\mathbf{W} = [W_1, \cdots, W_m]$ obtained by the GRASP represents the memory between the network's input layer and the network's hidden layer, where each $W_l = (w_{l1}, w_{l2}, \cdots, w_{l,n+1})^t, (l = 1, \cdots, m)$ determines a separating hyperplane.

4.2 The Pseudocode

```
Input: Training set {(X_i, d_i)}, (i = 1,...T)
Formulate problems (P2) y (P3)
Choose r = random{1,2}
Initialize Pat_rest1 = T_1 ; Pat_rest2 = T_2
While (Pat_rest1 > 0 and Pat_rest2 > 0)
   If r = 1
    Solve (P2), fixing all possible z_i = 0, one at
    a time
    Create RCL1
    If RCL1 ≠ {Ø},
       Choose r_1
       Drop patterns
       Update Pat_rest1
       Modify (P2) y (P3)
       r ← 2
   Otherwise (RCL1 = {Ø})
       Solve (P3), fixing all possible z_i = 0, one
       at a time
       Create RCL2
       Choose r_2
       Drop patterns
       Update Pat_rest2
       Modify (P2) y (P3)
 Otherwise (r = 2)
   Solve (P3), fixing all possible z_i = 0, one at
   a time
   Create RCL2
```

```
If RCL2 ≠ {∅},
    Choose r₂
    Drop patterns
    Update Pat_rest2
    Modify (P2) y (P3)
    r ← 1
Otherwise (RCL2 = {∅})
    Solve P2, fixing all possible z_i = 0, one at
    a time
    Create RCL1
    Choose r₁
    Drop patterns
    Update Pat_rest1
    Modify (P2) y (P3)
```

4.3 Completion of the Neural Network Design

If a neural network is constructed with the hyperplanes obtained by the GRASP and activation functions:

$$f(net_{ij}) = \operatorname{sgn}(net_{ij}),\ \ i = 1,\ldots,T\ ;\ j = 1,\cdots,m$$

as shown in Figure 10-1, then a mapping is obtained from the original set of patterns $X_i = (x_{i1}, x_{i2}, \cdots, x_{in})^t \in R^n, (i = 1,\ldots,T)$ into a set of images $Y_i = (y_{i1}, y_{i2}, \cdots, y_{im})^t \in R^m, (i = 1,\ldots,T)$ which is linearly separable.

To conclude the design of the network it is still necessary to determine the memory between the hidden layer and the output layer.

To do this, a training set $\{(Y_1, d_1), (Y_2, d_2), \cdots, (Y_T, d_T)\}$ will be created from the set of images $Y_i, (i = 1,\ldots,T)$ and solving a (P2)-type problem for this training set, a separating hyperplane will be obtained. With this last hyperplane it is possible now to complete the neural network that correctly classifies the original set of patterns in two classes or categories, as illustrated in Figure 10-2.

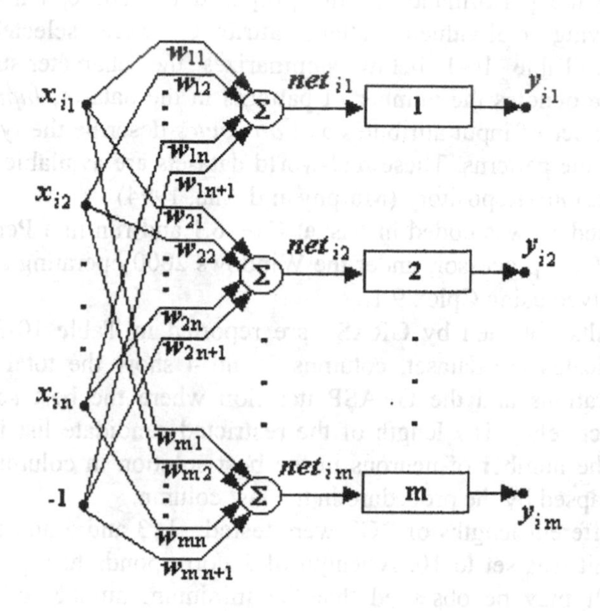

Figure 10-1. Hidden layer of the network.

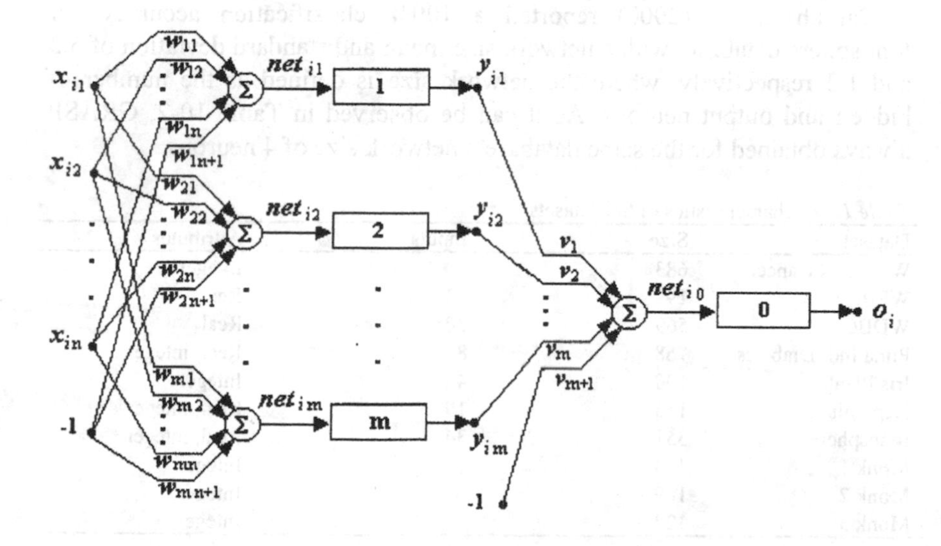

Figure 10-2. Neural network.

5. COMPUTATIONAL RESULTS

To show the performance of the proposed procedure, ten benchmark datasets having real-valued pattern attributes were selected for the experiments. Table 10-1 below summarizes the characteristics of the datasets. *Size* denotes the number of patterns in the dataset, *Inputs* indicate the total number of input attributes and *attributes* describe the type of input attributes of the patterns. These real-world datasets are available at the UCI Machine Learning Repository (Murphy and Aha, 1994) .

The procedure was coded in Visual C++ 6.1 and run in a Pentium 4 PC with a 3.00 GHz processor, under the Windows 2000 operating system. The LPs were solved using Cplex 9.1.

The results obtained by GRASP are reported in Table 10-2. The first column indicates the dataset, columns 2 and 4 show the total number of GRASP iterations and the GRASP iteration where the best solution was found respectively. The length of the restricted candidate list is shown in column 3, the number of neurons in the best solution in column 5 and the total time elapsed by the procedure in the last column.

Three different lengths of RCL were tested: 1, 3 and 5 and the GRASP iteration limit was set to 10. A length of 1 corresponds to a purely greedy procedure. It may be observed that the minimum number of neurons is attained always when the length of the RCL is set to 3, and the iteration limit to 10. It is worth to remark that the procedure is fast and the classification accuracy is always 100%, which is not the case in most reported works.

Parekh et al. (2000) reported a 100% classification accuracy for Ionosphere database, with a network size mean and standard deviation of 5.3 and 1.3 respectively, where the network size is defined as the number of hidden and output neurons. As it can be observed in Table 10-2, GRASP always obtained for the same database a network size of 4 neurons.

Table 10-1. Characteristics of the datasets

Dataset	Size	Inputs	Attributes
W. Breast Cancer	683	9	Integer
WPBC	198	32	Real
WDBC	569	30	Real
Pima Ind. Diabetes	768	8	Real, integer
Iris Plants	100	4	Integer
Hepatitis	155	19	Real, integer
Ionosphere	351	34	Real, integer
Monk 1	124	6	Integer
Monk 2	169	6	Integer
Monk 3	122	6	Integer

Table 10-2. Results of the GRASP runs

Dataset	# Iter.	RCL	Best Iter.	# neurons	Time(sec)
Breast	1	1	1	6	0
	5	3	2	5	32.031
	5	5	1	6	46.453
	10	3	2	5	90.171
	10	5	1	6	129.016
WPBC	1	1	1	5	0
	5	3	1	5	23.453
	5	5	3	5	19.124
	10	3	1	5	50.846
	10	5	3	5	44.936
WDBC	1	1	1	4	0
	5	3	2	4	100.78
	5	5	1	4	74.406
	10	3	7	3	197.298
	10	5	8	3	168.438
Iris Plants	1	1	1	3	0
	5	3	2	3	0.844
	5	5	1	3	0.641
	10	3	2	3	1.64
	10	5	1	3	1.5
Hepatitis	1	1	1	4	0
	5	3	1	3	1.594
	5	5	2	3	1.502
	10	3	1	3	3.703
	10	5	2	3	3.563
Pima	1	1	1	36	94.276
	5	3	1	37	974.734
	5	5	1	37	778.829
	10	3	7	36	1606.532
	10	5	1	37	1886.671
Ionosphere	1	1	1	3	0
	5	3	1	3	8.891
	5	5	1	3	14.687
	10	3	1	3	28.297
	10	5	1	3	35.734
Monk1	1	1	1	3	0
	5	3	4	4	1.75
	5	5	4	5	2.016
	10	3	8	3	2.829
	10	5	8	3	5.344

Dataset	# Iter.	RCL	Best Iter.	# neurons	Time(sec)
Monk2	1	1	1	8	0
	5	3	3	9	5.564
	5	5	4	9	5.763
	10	3	8	8	12.875
	10	5	10	9	15.719
Monk3	1	1	1	5	0
	5	3	2	4	1.516
	5	5	1	5	1.968
	10	3	6	3	3.281
	10	5	6	4	3.796

6. CONCLUSIONS AND FUTURE RESEARCHS

In this chapter we have approached the problem of designing a neural network in a different way from which it has been reported in the specialized literature. To the best of our knowledge, it is the first time that GRASP has been used to design a neural network for classification in two classes.

From linear formulations to determine separating hyperplanes, a GRASP was designed that computes the number of neurons in the hidden layer to solve the classification problem. In each iteration of the constructive phase, this algorithm determines one hyperplane, which separates patterns of a same class and stops when a linearly separable subset has been found. Moreover, if the set of patterns is linearly separable, the algorithm stops in the first iteration yielding the solution to the problem. Then, to finish the network design, the memory between the hidden layer and the output layer is determined solving a linear problem.

The proposed procedure reaches always a 100% classification accuracy and the execution time is good enough. In general, other constructive algorithms report a classification accuracy less than 100% and a greater number of neurons in the hidden layer when a 100% accuracy is reached. Thus, it can be said that the proposed algorithm can compete in quality and speed with other approaches.

This work can be generalized in order to consider classification problems in several categories.

ACKNOWLEDGEMENTS

This work has been partially supported by Conacyt under grant 36669A and is part of a Research Chair in Industrial Engineering of ITESM titled "Extended Enterprises for Mass Customization". The authors wish to acknowledge the support of these grants in the preparation of this manuscript.

REFERENCES

Angel-Bello F., Martínez J. L., and Alvarez A., 2003, A linear programming-based heuristic method for designing and training neural networks with discrete activation function for pattern classification, in: *Proceedings of the 8th International Conference on Industrial Engineering-Theory, Applications and Practice*, Las Vegas, pp. 748-753.

Cybenko, G. 1989. Approximation by superpositions of a sigmoidal function. *Mathematics of Control, Signals and Systems* **2**:303-314.

Feo, T. A., and Resende M. G. C., 1995, Greedy randomized adaptive search procedures. *Journal of Global Optimization* **6**:109-133.

Festa P., and Resende M., 2002, GRASP: An annotated bibliography , in: *Essays and Surveys in Metaheuristics,* C. C. Ribeiro and P. Hansen, eds., Kluwer Academic Publisher, Boston, pp. 325-367.

Flectcher J., and Obradovic Z., 1994, Constructive learning a near-minimal neural network architecture, *Neural networks* **1**:204-208.

Gori, M., and Tesi, A., 1992, On the problem of local minima in backpropagation, *IEEE Transactions on Pattern Analysis and Machine Intelligence* **14**:76-86.

Hornik, K., Stinchcombe, M., and White, H., 1989, Multilayer feedforward network are universal approximators, *Neural Networks* **2**:359-366.

Murphy P., and Aha D., 1994, Repository of machine learning databases; http://www.ics.uci.edu/AI/ML/MLDBRepository.html

Parekh, R., Yang, J., and Honavar, V., 2000, Constructive neural network learning algorithms for pattern classification, *IEEE Transactions on Neural Networks* **2**:436-451.

Roychowdhury, V., Siu, K.-Y., and Kailath, T., 1995, Classification of linearly nonseparable patterns by linear threshold elements, *IEEE Transactions on Neural Networks* **2**:318-331.

Yu, X., 1992, Can backpropagation error surface not have local minima?, *IEEE Transactions on Neural Networks* **3**:1019-1021.

Chapter 11

MEMETIC ALGORITHMS

Natalio Krasnogor[1], Alberto Aragón[2] and Joaquín Pacheco[2]
[1] *School of Computer Science and I.T. University of Nottingham. England;*
[2] *Departamento Economia Aplicada. University of Burgos, Spain*

Abstract: This chapter introduces and analyzes a memetic algorithm approach for the training of artificial neural networks, more specifically multilayer perceptrons. Our memetic algorithm is proposed as an alternative to gradient search methods, such as *backpropagation*, which have shown limitations when dealing with rugged landscapes with many poor local optimae. The aim of our work is to design a training strategy that is able to cope with difficult error manyfolds, and to quickly deliver trained neural networks that produce small errors. A method such as the one we proposed might also be used as an "on-line" training strategy.

Key words: Memetic algorithms; neural network; metaheuristic algorithms; evolutionary algorithms.

1. INTRODUCTION TO MEMETIC ALGORITHMS

Memetic Algorithms (MAs) are a class of stochastic global search heuristics in which evolutionary algorithms-based approaches are combined with problem-specific solvers. The later might be implemented as local search heuristics techniques, approximation algorithms or, sometimes, even (partial) exact methods. The hybridisation is meant to either accelerate the discovery of good solutions, for which evolution alone would take too long to discover, or to reach solutions that would otherwise be unreachable by evolution or a local method alone. As the large majority of memetic algorithms use heuristic local searches rather than, e.g., approximation or exact methods, in what follows we will focus only on the local search add-ins for the Evolutionary Algorithm (EA). It is assumed that the evolutionary search provides for a wide exploration of the search space while the local search can somehow zoom-in on the basin of attraction of promising solutions. MAs have been proven very successful across a wide range of domains such as combinatorial optimization (Merz, 2000), optimization of non-stationary functions (Vavak et.al, 1996), multi-objective optimization (Knowles and Corne, 2001), bioinformatics (Krasnogor, 2004), etc.

Memetic algorithms have received various names throughout the literature and scientist not always agree what is and what is not an MA due to the large variety of implementations available. Some of the alternative names used for this search framework are hybrid GAs, Baldwinian EAs, Lamarckian EAs, genetic local search algorithms, and other names are not unheard of. (Moscato 1989) coined the name memetic algorithm to cover a wide range of techniques where evolutionary-based search is augmented by the addition of one or more phases of local search.

The natural analogies between human evolution and learning, and EAs and artificial neural networks (ANNs) prompted a great deal of research into the use of MAs to evolve the design of ANNs. Some research concentrated mainly in the automated design of the architecture of interconnection among neurons, which is a combinatorial optimisation problem, and others on the adjustment of the weights associated to the links between neurons, which is a continuous optimisation problem. During the 1980s and early 1990s, ANNs were trained using, for example, back-propagation, conjugate gradients or related methods. At the same time, seminal work by Hinton and Nowlan in the late 80s (Hinton and Nowlan 1987) provided much insights into the interplay between evolution and learning. Other researchers (Whitley et al. 1994; Mayaley 1996; Turney 1996 and Houck et al. 1997) followed similar trends, which reinforced the perception that, in order to distil an evolutionary algorithm that could achieve maximum performance on a real world application, much domain knowledge needs to be incorporated. Domain knowledge was oftentimes encoded by means of problem specific local searchers.

Research in memetic algorithms has progressed substantially, and several Ph.D. dissertations have been written analysing this search framework and proposing various extensions to it (Hart 1994; Land 1998; Merz 2000; Moscato 2001 and Krasnogor 2002). There are several reasons why it is worthwhile hybridising evolutionary algorithms with local searchers, among them we could mention:

1. Complex problems might be decomposable to some extent and its not far fetched to think that different subproblems could be better solved by different methods.
2. The EA (or the local searcher) can be used as pre/post processors of solutions. One often uses a local searcher in tandem with an evolutionary search as a means of fine-tuning or repairing the best

solution(s) produced by the evolutionary algorithm. In turn, one might have very powerful local search methods but would like to introduce diversity (or robustness) into the solutions afforded by the local searcher, in that case one could use an evolutionary algorithm after several runs of local search have delivered the initial population.

3. (Sub)Problem specific information can be distilled into variation operators, e.g. crossover and mutation, or into local searchers as to effectively bias the search process towards promising regions of the search space.
4. In some cases there are exact/approximate methods for (some of) the subproblems and, when these are available, they can be incorporated into the evolutionary algorithm.
5. (Sub)Problems have constraints associated to solutions and heuristics/local search are used to repair solutions found by the EA. If heuristic/local search strategies in MAs are considered "first class citizens", rather than simple plug-ins, then a much richer definition of adaptive hybrid metaheuristics is possible: the local search strategies might be generated at the *same time* with the solutions they intend to improve (e.g. see Krasnogor 2004)

However, the reader must note that it is well established that generally good black-box optimisers that excel at solving any problem do not exist. This is why successful EAs are usually found in "hybridized form". Thus, a "conservation of competence" principle applies: the better one algorithm is solving one specific instance (class) the worst it is solving a different instance (class) (Wolpert and Macready 1997). That is, it cannot be expected that a black-box metaheuristic will suit all problem classes and instances all the time, because, it is theoretically impossible to have both **ready made of-the-shelf general and good** solvers. Memetic algorithms are popular because they are good a *algorithmic template* that aid in the balancing act needed to solve challenging problems by successfully re-using a general, of-the-shelf, solver (e.g. EAs), which can readily incorporate domain specific features in the form of local searchers.

Virtually everybody agrees that for an evolutionary algorithm to succeed in a real-world scenario it must contain sufficient domain knowledge, which is built-in not only in the representation, fitness function and genetic operators used but also in the local search(ers) employed. However, researchers and practitioners sometimes fail to consider in detail some of the main design issues that are pertinent to memetic algorithms. In the next section we will assume that the reader is familiar with both evolutionary algorithms and local search and we will address some important issues that must be kept in mind when designing MAs.

1.1 Evolutionary Algorithms + Local Search = Memetic Algorithms

As suggested above, there are a number of benefits that can be gained by combining the global search of EAs with local search or other methods for improving and refining an individual's solution. However, as there are no free lunches these benefits must be balanced against an increase in the complexity in the design of the algorithm. That is, a careful consideration must be place on exactly *how* the hybridisation will be done.

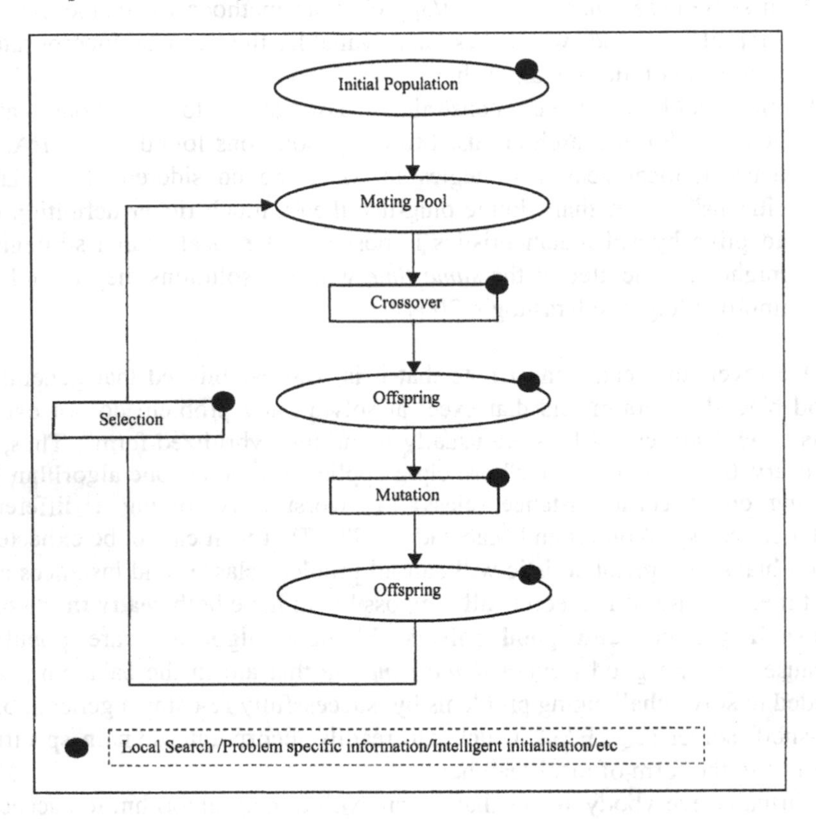

Figure 11-1. A Memetic Algorithm Template.

Consider for example the Memetic Algorithm template in Figure 11-1. A quick glance at the figure allows us to identify the basic structure of an evolutionary algorithm for which *hot-spots,* i.e the places where hybridisation could take place, have been identified and marked with red circles. Each of these hot-spots provides opportunity for hybridisation. For example, the initial population could be seeded with solutions arising from sophisticated problem specific heuristics, the crossover (mutation) operator could be enhanced with domain specific and representation specific constrains as to provide better searchability to the EA. Moreover, local search could be applied to any or all of the intermediate sets of solutions (e.g. the offspring set). Hard won empirical evidence points to the fact that the more problem-specific knowledge is incorporated within a memetic algorithm the better it will perform. However, the most popular form of hybridisation is to apply one or more phases of local search, based on some probability parameter, to individual members of the population in each generation. Some, but not all, of the most important design decisions one needs to make when designing an MA, are concerned with one or more of the following: Lamarckianism vs. Baldwinian effect, preservation of diversity vs. the need to quickly zoom-in on good solutions, the choose of neighbourhoods for the local searcher and use of performance profiles. In what follows we briefly describe each of them.

1.2 Lamarckianism vs. Baldwinian Effect

When integrating local search with evolutionary search we are faced with the dilemma of what to do with the improved solution that is produced by the local search. That is, suppose that individual i belongs to the population P in generation t and that the fitness of i is $f(i)$. Furthermore, suppose that the local search produces a new individual i' with $f(i') < f(i)$ for a minimisation problem. The designer of the algorithm must now choose between two alternative options. Either (option 1) he/she replaces i with i', in which case $P = P - \{i\} + \{i'\}$ and the genetic information in i is lost and replaced with that of i', or (option 2) the genetic information of i is kept but its fitness altered: $f(i)= f(i')$. The first option is commonly known as Lamarckian learning while the second option is referred to as Baldwinian Learning (Baldwin, 1896). The issue of whether natural evolution was Lamarckian or Baldwinian was hotly debated in the nineteenth century until Baldwin suggested a very plausible mechanism whereby evolutionary progress can be guided towards favorable adaptation without the inheritance of life-time acquired features. Unlike in natural systems, the designer of a memetic algorithm may want to use either of these adaptation mechanisms. Hinton and Nowlan (1987) showed that the Baldwin effect could be used to improve the evolution of artificial neural networks, and a number of researchers have studied the relative benefits of Baldwinian versus Lamarckian algorithms, e.g., Whitley et al. (1994), Mayaley (1996), Turney (1996), Houck et al. (1997), etc. Most recent work, however, favored either a fully Lamarckian approach, or a stochastic combination of the two methods. It is a priori difficult to decide what method is best, and probably no one is better in all cases. Lamarckianism tends to substantially accelerated the evolutionary process with the caveat that it often results in premature convergence. On the other hand, Baldwinian learning is more unlikely to bring a diversity crisis within the population but it tends to be much slower than Lamarckianism.

1.3 Preservation of Diversity

We have mentioned above the difficulties one may face when using a very aggressive local search method, chief among those difficulties is the need to preserve diversity. Diversity could be lost if, for example, every individual within the population would be taken to a local optimum by means of a vigorous local search (i.e. complete local searches), or, in the case of partial local searches, if the search space under consideration had very wide basins of attraction from which crossover and mutation could not easily escape. Various mechanism have been studied as a way to avoid premature convergence in Memetic Algorithms. For example, if a smart population initialisation is performed, only a small proportion of individuals should be seeded with problem specific information. Alternatively, other authors have resorted to do local search selectively, that is, only a small number of individuals are improved by local search. A very common strategy to deal with diversity preservation has been the introduction of very specific crossover operators (e.g. Freisleben and Merz 1996) that help keeping the genetic variety always above a minimum threshold. The modification of the selection operator as to prevent multiple copies of individuals has also been a preferred strategy (Eshelman 1990). More recently we have introduced three distinct and powerful method which not only help preserve diversity but also enhance the overall algorithmic performance. Multiple local searchers, where each one induces a different search space with distinct local optima thus avoiding local traps have been used in (Krasnogor and Smith 2001; Krasnogor et.al. 2002 and Krasnogor 2004). Also, fuzzy sets and systems have been implemented to explicitly control diversity within the decision rule in the local search stage (eg. Krasnogor and Pelta 2002). Finally, it is also possible to modify the selection operator, or local search acceptance criteria, to use an Adaptive Boltzmann method so as to preserve diversity (eg. Krasnogor and Smith 2000). This is similar to simulated annealing (Kirckpatrick et.al 1983), where worsening moves can be accepted with nonzero probability to aid escaping from local optima. A promising method that tackles the diversity issue explicitly was proposed in Krasnogor and Smith (2000), where during the local search phase a less-fit neighbor may be accepted with a probability that increases exponentially as the range of fitness values in the population decreases:

$$prob(accept) = \begin{cases} 1 \Leftrightarrow \Delta f > 0 \\ e^{k^* \frac{\Delta f}{f_{max} - f_{avg}}} & otherwise \end{cases}$$

where k is a normalisation factor, $\Delta f = f(i') - f(i)$, and we are assuming a maximisation problem. This mechanism seamlessly induces the Memetic Algorithm to oscillate between periods of intense exploitation whenever the spread of fitness in the population, i.e $f_{\max} - f_{avg}$, is large, and periods of vigorous exploration when that spread is confined to a narrow interval.

1.4 Which Neighborhood Must the Local Searcher Use?

When the researcher is confronted with the design of a memetic algorithm he must carefully consider the topology he/she will give to the space of all feasible solutions. In the case of local search, this is done by defining the structure of the neighborhood of a solution, that is, by specifying which are the feasible solutions that are reachable from a given point in the solution space. If the neighborhood is too small or restrictive it is unlikely that good local optima will be outputted by the local search. On the other hand, if the neighborhood is complete, then from every point in the search space it is possible to build a path to any other point, but in this case, completeness is paid with very long (maybe even exponential) searches. Furthermore, the size of the neighborhoods –although important– is not the only feature to consider. Merz and Freisleben (1999), Kallel et al (2001) and many others have shown through some statistical analysis of fitness landscapes that the choice of move operator can have a significant impact on the efficiency and effectiveness of the local search, and hence of the overall performance of the Memetic Algorithm. Moreover, we have formally shown recently that, in order to reduce the worst-case run times of evolutionary algorithms hybridised with local search, it is necessary to choose a local search method whose move operator is different from the ones that the recombination and mutation operators (Krasnogor, 2002) induce. This formalizes the intuitive point that within an algorithmic template such as a memetic algorithm, the recombination, and particularly the mutation phase, have valuable roles in generating points that lie in different basins of attraction with respect to the local search operator. This last result should be of particular interest to the practitioner as it is usually easy to provide the memetic algorithm with several distinct local searchers and endow it with the capacity to learn, on-the-fly, which one to use (Krasnogor and Smith 2001; Krasnogor et al 2002; Krasnogor 2004 and Krasnogor and Gustafson 2004). The use of a set of possible local search strategies is analogous to Dawkin's memes (Dawkins 1976). The extension of this approach to allow the adaptation of the local search "memes'" in the form of a coevolving population, and the implications for search is currently under way in different research groups most notably by J.E. Smith at the University of the West of England, Bristol and N. Krasnogor at the University of Nottingham, both in the U.K.

1.5 Use of Performance Profiles

Performance profiles could be use to track the progress of the search vis-à-vis the various algorithmic components of the memetic algorithm, that is, one might want to take into consideration the reuse of knowledge gained about the search landscape(s) the algorithm is exploring *throughout* the optimisation process. One possible hybridization that explicitly uses knowledge about previously seen solutions as a mean to guide optimisation is with tabu search (Glover 1989). For example, the "tabu" list would maintain a set of already visited points, which the algorithm is forbidden to re-visit thus also helping to maintain diversity. In a related way, the adaptive Boltzmann scheme mentioned above could be extended as to keep track not only of the spread of fitness within the current generation but more generally of the genotypic variety on historical populations.

1.6 Specific Considerations for Continuous Domains

The design issues we mentioned above are mainly concerned with the optimisation of combinatorial, i.e. discrete, problems, such as the optimisation of the architecture of a neural network. As previously mentioned, the automated design of neural networks also involves the optimisation of continuous problems, e.g., weight adjustments. In this case, a person engineering a memetic algorithm, must mull over different design factors. As it is well known, optimisation in continuous domains requires that appropriate search scales be identified for both the local and global search. Moreover, it is often difficult to recognise whether a feasible solution represents or not a local optimum. If gradient information is not available then very long local searches might be needed to ensure convergence with large accuracy. The uncertainly about which local search method is better on which problems is even more acute in continuous optimization. Many different local search methods have been proposed, but as they are general methods it is not clear whether any given local search method is effective for a particular continuous optimisation task. Thus the design of competent MAs for continuous domains can be quite different than for combinatorial problems. A simple example will clarify what we mean: it is common in memetic algorithms for combinatorial search (although not always desirable) to apply local search until local optimae are identified, however, in general one cannot assume that a local search method will quickly converge towards a good local optimum within a continuous domain. Researchers and practitioners working on continuous domains have found that this is a usual problem when applying derivative-free methods (e.g. the Nelder-Mead simplex), but it may also occur in light of derivative information. Two common strategies are normally used to deal with this difficulty, which essentially account for a careful balancing between global and local search. This balance has been often implemented as either truncated local searches or selective local searchers. The reader is invited to consult with (Rudolph, 1996; Hart et.al., 2003) for further insights on these issues.

Memetic algorithms are very powerful algorithmic templates that can be applied to a wide variety of problems and they have been gaining the favor of both researchers and practitioners as they provide a simple formula that allows to combine a robust global search technique, i.e. evolutionary algorithms, with powerful, domain-specific local searchers. The reader interested in a more detailed discussion (with a large variety of examples) is refered to Hart, Krasnogor and Smith (2004) and Krasnogor and Smith (2005).

2. NEURAL NETWORK ARCHITECTURE

In what follows we describe a training problem for multi-layer perceptrons and the memetic algorithm we have used to tackle it. The architecture of the neural network is assumed to be fixed once the training starts.

Consider a multiple-layer perceptron neural network with just one hidden layer. Let n denote the number of neurons in the input layer (which is logically the same as the number of input variables), and m the number of neurons in the hidden layer. Let us assume that the network has only one neuron in the output layer. Each neuron in the hidden layer, as well as in the output layer, has a bias term which functions as a constant. A schematic representation is shown in Figure 11-2 (from Laguna and Martí 2002).

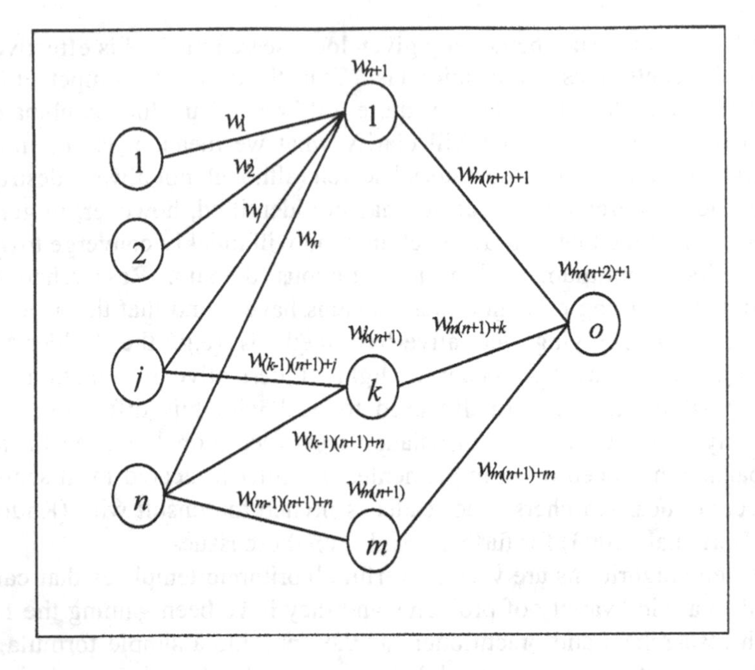

Figure 11-2. Neural network architecture.

Weights are sequentially numbered such that the weights that link the input layer with the first neuron of the hidden layer range from w_1 to w_n where their bias term is w_{n+1}.

The network reads an input vector $e = (e_j)$ and the information propagates from one layer to another until output s is produced. Thus, the input information of each neuron, $k = 1, .., m$, of the hidden layer is given by

$$z_k = w_{k(n+1)} + \sum_{j=1}^{n} w_{(k-1)(n+1)+j}\, e_j$$

and the output neuron receives the following data

$$w_{m(n+2)+1} + \sum_{k=1}^{m} w_{m(n+1)+k}\, a_k$$

where a_k are outputs from the hidden neurons.

In our example we use a sigmoid function as a transfer function for the hidden layer and an identity function for the output neuron. Thus,

$$a_k = \frac{1}{1+e^{-z_k}} \quad \text{and}$$

$$s = w_{m(n+2)+1} + \sum_{k=1}^{m} w_{m(n+1)+k}\ a_k$$

3. THE TRAINING PROBLEM

3.1 Background

Training consists in adjusting the weight vector $w = (w_p)_{p=1..m(n+2)+1}$, in the following way: a set of training vectors S is presented, each consisting of an input and output vector; $\forall\ (e, y) \in S$, e is the input vector and y the expected output. Training consists of determining the weights that minimize the difference between the expected output and those produced by the network o when given the input vectors e. Using the least squares criterion this can be expressed as

$$\min_w E = \frac{1}{|S|} \sum_{(e,y)\in S} (o-y)^2$$

where E is the Mean Squared Error.

The values of each of the variables of the input vectors are normalized between -1 (the lowest value in the training set) and +1 (the highest) for the training stage; the output values are also normalized between –0.8 and +0.8. This normalization is recommended in different works, (Sexton et al., 1998 and 1999).

The algorithms based on gradient descent, such as the *back-propagation method* (developed by Werbos 1974 and later by Parker 1985; LeCun 1986 and Rumelhart et al. 1986), basically use the following formula to update the weights

$$w_{ji} = w_{ji} - \eta \frac{\partial E}{\partial w_{ji}}$$

where η is known as the *learning parameter*. This is an iterative process that terminates when there is no further improvement in E. In fact, one of the limitations of this method is convergence to poor local optimae and their dependence on the initial solution. In addition, they are not applicable to network models that use transfer functions for which derivatives cannot be calculated.

The most recent global optimization methods, especially metaheuristic strategies, have generally obtained better results than previous methods based on gradient descent. Thus, traditionally, *genetic algorithms* have been highlighted in works such as Montana and Davis (1989), Schaffer et al. (1992), Schaffer (1994) and Dorsey et al. (1994). *Memetic algorithms* have also yielded interesting results as in the works of Moscato (1993), Topchy et al. (1996), Yao (1993) and Ichimura and Kuriyama (1998). Excellent results have also been obtained in Sexton et al. (1998) and (1999) with *genetic algorithms*, *tabu search* and *simulated annealing*, and in Laguna and Martí (2002), where *scatter search* was used.

The aims of our paper is to propose learning strategies that:

1. Can minimise the mean squared error E for the training set (i.e. learning)
2. Can minimise the mean square error E for unseen input sets (i.e. generalization)
3. Are fast and robust across a variety of domains and that can be used as on-line learning methodologies.

Points 1 and 2 above refer to the well known trade-off between strong memorisation and good generalization. We are aiming at producing a learning protocol that delivers the correct balance of both. On the other hand, we want our protocols to be fast and robust but at the same time scalable. By scalable we mean that if the training method is given more CPU time to learn then it can minimise even further the error term E. In adition we would like that these new strategies are consistent, i.e., that they are not too sensitive to initial conditions such as the particular training set used or the initial population of the memetic algorithm.

3.2 The Memetic Algorithm

We describe next a memetic algorithm for the training of neural networks together with its main characteristics. One aspect to take into account is that due to the type of transfer functions used, these methods only seek weight values in the hidden layer: those in the output layer are obtained by the least squares method, ensuring good results, (this was suggested by Sexton et al., 1998). In Section 4 we describe an improvement method used by the memetic algorithm. The memetic algorithm is described in Section 5.

4. IMPROVEMENT BY LOCAL SEARCH

In this section we analyze the probabilities of improving a given solution by using a simple hillclimbing algorithm. The hillclimber algorithm starts from an initial solution (i.e. a weights vector *w*), samples a small set of weights vectors around *w*, and selects as its next vector the one that minimises the error *E*. Specifically, let **w** be any solution with w_l components and *radius* a sufficiently small real number; the following procedures are defined:

Change_Weight Procedure (***w***, ***radius***, ***w'***)
  *Generate a solution **w'** whose components w_l' take random values in the interval (w_l – radius, w_l + radius)*

Likewise, given that $\mathbf{w_0}$ is any solution of the parameters *n_neighbours* ∈ *N* and *radius* ∈ *R*; the following is defined:

Best_Neighbour Procedure (***w₀***, ***radius***, ***n_neighbours***, ***w'***)
  min = ∞
  For i = 1 up to n_neighbours do:
    *Change_Weight (**w**, radius, **w''**)*
    *If E(**w''**) < min do: **w'** = **w''** and min = E(**w''**)*

Finally, given that **w** is an initial solution we define

Best_Environment Procedure (***w***, ***radius***, ***n_neighbours***, ***w'***)
  Repeat
    *Previous_value = E(**w**)*
    *Execute Best_Neighbour (**w**, radius, n_neighbours, **w'**)*
    *If E(**w'**) < E(**w**) do **w** = **w'***
  *Until E(**w'**) >= Previous_value*

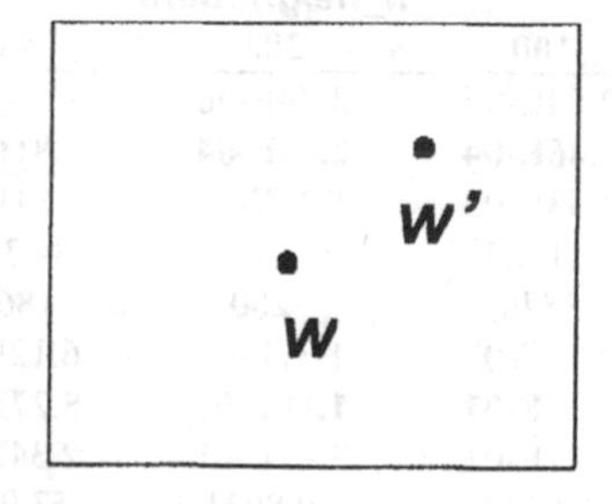

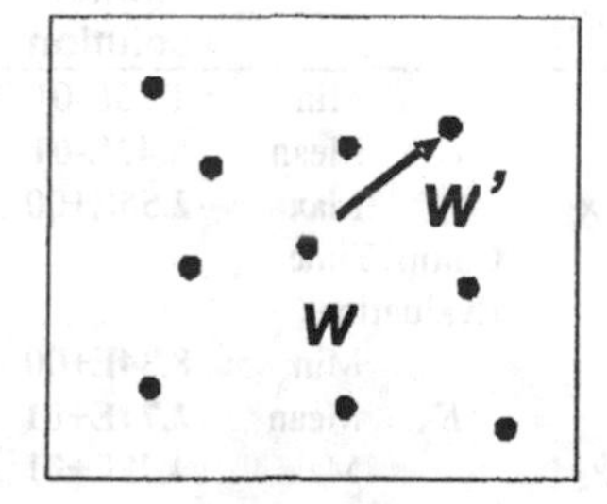

Figure 11-3. The Change_Weight and Improve_Neighbour procedures.

The latter procedure is thus an improvement method: in each iteration a set of points is generated by changes around the current solution. Then, the best solution is determined and if this one improves the current solution the change is implemented. The process stops when, from among the solutions generated in one iteration, none is found which improves the current one.

4.1 Analysis of the n_neighbours and the Radius Parameters

To study the influence of the number of neighbour solutions generated in each pass, a set of 50 training vectors with 2 input variables x_1 and x_2 is generated. The input values are integers generated within [-100, +100] for x_1 and within [-10, +10] for x_2. The output values y are given by $y = x_1+x_2$. In addition, a second set of training vectors is generated with the same input vectors and $y = x_1 \cdot x_2$ as the output vectors. Similarly, 10 initial solutions (hidden layer weights) are generated for each problem. Their components are random values between –5 and +5. With each solution as the starting point the *Best_Environment* procedure described earlier was executed. Different values of *n_neighbours* were used: 100, 200 and 500. In all cases the value of *radius* takes 0.1. Table 11-1 shows the following results for each problem: mean computational time in seconds, number of solutions generated during each process, and statistical values obtained for E (minimum, mean and maximum). All tests were done on a 600 Mhz Pentium III (DUAL processor). The BORLAND PASCAL (7.0) and BORLAND DELPHI (3.0) compilers were used.

Table 11-1. Hillclimber analysis

Function			Initial solution	*n_neighbours* 100	200	500
		Min	1.25E-01	2.64E-06	3.74E-06	1.55E-06
	E	Mean	**8.42E-01**	**4.46E-04**	**2.74E-04**	**9.81E-05**
$y = x_1 + x_2$		Max	2.55E+00	1.33E-03	1.33E-03	8.81E-03
	Comp. Time			7.1283	13.2654	34.3502
	Evaluations			3870	7200	18650
		Min	8.34E+00	3.10E-02	1.37E-02	6.12E-03
	E	Mean	**2.74E+01**	**1.97E-01**	**1.19E-01**	**5.27E-02**
$Y = x_1 \cdot x_2$		Max	4.72E+01	5.91E-01	3.82E-01	2.34E-01
	Comp. Time			9.5344	19.8031	53.9314
	Evaluations			5140	10680	29100

As expected, a larger number of generated neighbours improves the final result, although at the cost of greater computational time.

To analyse the influence of the radius used to generate values around the current solution, the same tests were carried out, i.e. the same sets of training vectors and 10 initial solutions generated for each of them. The following values were used for the parameter *radius*: 1, 0.1 and 0.01. To avoid excessive computational time *n_neighbours* was set to 100. The results are shown below:

Table 11-2. Radius analysis

Function			Initial solution	Radius 1	Radius 0.1	Radius 0.01
$y = x_1 + x_2$	*E*	Min	2.15E-01	5.33E-05	9.62E-05	7.13E-08
	E	Mean	**8.14E-01**	**6.66E-03**	**1.01E-03**	**2.10E-04**
	E	Max	1.29E+00	3.19E-02	2.20E-03	9.99E-04
	Comp. Time			0.8517	5.3984	55.9828
	Evaluations			440	2790	28940
$y = x_1 \cdot x_2$	*E*	Min	7.23E+00	1.82E-01	1.15E-02	1.12E-03
	E	Mean	**1.68E+01**	**8.62E-01**	**1.34E+00**	**2.93E-01**
	E	Max	2.56E+01	2.30E+00	1.07E+01	1.55E+00
	Comp. Time			1.3703	7.464	90.8813
	Evaluations			690	3760	45780

Decreasing the *radius* greatly decreases the error, but at the cost of increasing the computational time. Generally, high values of *radius* entail a rapid decrease in computational time whereas low values lead to slower behaviour but yield better solutions in the long run. In addition, it is clear that the effect of the parameter *radius* is more significant than that of *n_neighbours.*

4.2 Improvement with a Variable Radius

As mentioned in previous sections, our method attempts to combine speed and "depth" (i.e. quality of the final solution). This method, called *Variable_Radius_Improvement*, consists in modifying the algorithm such that the value of *radius* changes depending on whether improvements are found or not: if the hillclimber finds a better solution, the new solution is accepted and the value of the *radius* with which the hillclimber searched is increased. On the other hand, if no improvement is found the value of the radius is decreased and takes effect in the following iterations. The algorithm ends when a given number of consecutive iterations have taken place without any improvement. That idea has been taken form Vavak, Fogarty and Jukes (1996), Krasnogor and Smith (2000) and Krasnogor and Pelta (2002).

With **w** as an initial solution, the real parameters *radius0* (initial value of *radius*) and *alpha*, and the integer parameters *n_neighbours* and *max_int*, the proposed algorithms can be written in pseudocode as follows:

Improve_variable_radius Procedure (*w, radius0, alpha, n_neighbours, max_int*)
Do radius = radius0, n_int = 0;
Repeat
Execute Best_Neighbour (w, radius, n_neighbours, w')
*If E(w') < E(w) do: w = w', radius = radius*alpha and n_int = 0*
Otherwise: radius = radius/alpha and n_int = n_int+1
Until n_int = max_int

To compare how well this procedure works, the same set of training vectors have been used as in subsection 3.1. Similarly, 10 initial solutions were generated for each of these sets. The parameters take the following values: *n_neighbours* = 10 and 25, *alpha* = 10, *radius0* = 1 and *max_int* = 3. The results were as follows:

Table 11-3. Comparing *n_neighbours* in *Improve_variable_radius*

Function			Initial solution	*n_neighbours* 10	25
	E	Min	4.22E-02	5.83E-06	3.44E-07
	E	Mean	**5.31E-01**	**2.41E-04**	**1.05E-04**
$y = x_1 + x_2$	*E*	Max	1.73E+00	1.15E-03	4.84E-04
	Comp. Time			3.4296	3.3905
	N. Evaluations			1850	1832
	E	Min	6.02E+00	3.23E-04	9.60E-05
	E	Mean	**1.52E+01**	**1.19E-01**	**2.71E-03**
$y = x_1 \cdot x_2$	*E*	Max	2.28E+01	9.89E-01	1.55E-02
	Comp. Time			27.0063	253.6608
	N. Evaluations			14651	137615

A comparison of Tables 11-1, 11-2 and 11-3 reveals very interesting results: good solutions are obtained within reasonable computational times. Therefore, in improvement by local search methods, an appropriate choice of radius, and the possibility of locally changing the radius is more decisive than increasing the number of neighbours.

In any case, the improvement methods proposed in this section are at least "universal", since they can easily be applied to any problem and are not specially made for the network model. For example, no derivability or continuity are demanded from the transfer functions, but this is not the case with gradient-based methods.

5. MEMETIC ALGORITHM

In the previous section we described and tested the hill-climber method with which we hybridised a genetic algorithm, thus obtaining a memetic approach. The pseudocode in Figure 11-4 shows how the genetic operators, local search method and selection procedure are "tied-up" within the MA.

The improvement procedure is *Variable_Radius_Improvement* with the same values as the parameters $radius0 = 0.1$, $alpha = 1$, $n\_neighbours = 2$, $max\_int=1$. The *fitness* of each individual is given by the value of the corresponding objective function E. The probability of selecting a given solution w is proportional to $Emax - E(w)$, where $Emax$ is the maximum value of $E(w)$ within the population.

The crossover operator is a "*one-point crossover*", that is, if $\mathbf{w}$ and $\mathbf{w'}$ are two parent solutions,

$$w = (w_1, w_2, .., w_{m(n+1)}) \quad \text{y} \quad w' = (w'_1, w'_2, .., w'_{m(n+1)})$$

a crossover point is randomly generated between 1 and $m(n+1)$-1, (crossover_point), in such a way that the offspring solutions are defined as

$$w^* = (w_1, w_2, .. , w_{point\_crossover}, w'_{point\_crossover+1}, .. , w'_{m(n+1)})$$
$$w^{**} = (w'_1, w'_2, .. , w'_{point\_crossover}, w_{point\_crossover+1}, .. , w_{m(n+1)})$$

Memetic Procedure:
Generate an initial population of solutions
Improve them with a pre-established method
Repeat
- *Create a mating pool by sampling the population with fitness proportional probability.*
- *Crossover: Randomly pair members (i.e. parents) from the mating pool and apply the crossover operator producing 2 offspring from each pair. Newly created offspring are added to the offspring set.*
- *Mutation: apply with low probability the mutation operator to the offspring set.*
- *Perform local search with the variable radius hill-climber starting from each member of the offspring set*
- *Replace the worst solutions in the population with the new offspring*
- *Implement the mutation parameter*

Until reaching some stopping criterion

Figure 11-4. Memetic Algorithm's pseudocode.

The weights of each new child solution can change or *mutate* with a low probability, *p_mut*. To decide whether a given weight changes or not, a random value is uniformly generated in (0, 1); if this value is less than *p_mut* the change is implemented. This consists in assigning to it a new random point in the interval (-0.5, +0.5, *mutation*). This mutation process aims at adding diversity to the process and prevents it from being locked in a poor local minimum.

A relatively original feature of our procedure is that the value of parameter *p_mut* is not fixed but varies between two pre-set values *p_mut_min* and *p_mut_max*. The aim is as follows: when a new and better solution is found the mutation probability decreases, and thus the search intensifies in such a region. As generations pass (iterations) without improvement, the mutation probability increases to intensify the search in other regions. Three rules govern the adaptation of the mutation rate:

- If in the current iteration a new and better solution is found do $p\_mut = p\_mut\_min$
- In the opposite case do $p\_mut = p\_mut + p\_mut\_inc$
- If $p\_mut > p\_mut\_max$ do $p\_mut = p\_mut\_max$.

The values *p_mut_min* and *p_mut_max* represent the maximum and minimum mutation probabilities.

The following values are used: $p\_mut\_min = 0{,}05$, $p\_mut\_max = 0{,}1$ and $p\_mut\_inc = 0{,}0005$. In addition, the number of population elements is 100, and the number of selected parents (size of *mating pool*) is 20. In each iteration the worst 20 elements are replaced by the new 20 offsprings.

6. COMPUTATIONAL RESULTS

In order to test the quality of the memetic algorithm, and compare it with other relevant methods found in the literature, we carried out the various tests. We have considered 6 test-beds:

1) $y = x_1 + x_2$

2) $y = x_1 x_2$

3) $y = \frac{x_1}{1+|x_2|}$

4) $y = x_1^2 + x_2^3$

5) $y = x_1^3 + x_2^2$

6) $y_t = y_{t-1} + 10.5\left(\frac{0.2y_{t-5}}{1+(y_{t-5})^{10}} - 0.1y_{t-1}\right)$

Note that these are continuous and derivable functions except for the third one. The sixth function is a discrete version of the well-known MacKey-Glass equation that has been frequently used in neural network research (Gallant and White 1992; Goffe et al. 1994).

For each of the first five functions a training set with 50 vectors is considered: the input values are integers generated within [-100, +100] for x_1 and within [-10, +10] for x_2; the output value y is given by the corresponding function. The same input vectors were used in the five cases. For the sixth function the five nodes or input variables were used although, obviously, only three were needed. However, their inclusion for checking the network's capacity to ignore unnecessary information is of interest. In this case, the training vectors were recursively generated from the values (1, 6, 0, 0, 0, 0).

The following methods were compared:

- BP: Backpropagation algorithm used in *Neural Works Profession II/Plus* from *NeuralWare*. This freeware package was rated the best among several commercial programs in Sexton et al. (1998).
- TS: *Tabu* Search algorithm proposed in Sexton et al. (1999).
- SA: Simulated *Annealing* algorithm proposed in Sexton et al. (1999).
- GA: The Genetic Algorithm proposed in Sexton et al. (1999).
- SS: *Scatter* Search Algorithm proposed by (2002).
- ME: *Memetic* Algorithm proposed in Section 5.

The tests were repeated ten times for each method and function. Table 11-4 shows the best results obtained by the different algorithms for each of these problems where we identified in bold the best result. The results of each method are taken from the references mentioned where the same functions and neural network architectures were used.

Table 11-4. Best results obtained for BP, SA, GA, TS and SS

Function	BP	SA	GA	TS	SS	ME
1	5.23E-01	1.05E-05	4.16E-07	1.42E-06	8.93E-08	**6.32E-14**
2	1.10E+01	3.17E-02	1.27E-02	8.38E-02	5.97E-03	**6.56E-05**
3	8.43E+00	1.76E+00	1.82E-01	2.45E-01	2.96E-03	**4.62E-04**
4	1.88E+02	4.84E-01	4.09E-02	4.32E-01	2.31E+00	**1.93E-03**
5	8.57E+03	4.39E-01	**3.06E-03**	2.56E-02	2.76E+02	4.58E-01
6	1.55E-01	1.02E-02	2.53E-02	5.35E-02	3.34E-03	**1.51E-04**

According to Laguna and Martí (2002) and Sexton (1998) and (1999), SS limits the number of evaluations of the objective function to 50,000; BP carries out 4.18 million, SA between 112,501 and 12.6 million, GA 100,000, TS between 190,021 and 928,061, and ME 100,000. Our ME achieves the best results in all functions except in the 5th function, where it is outperformed by SA, GA, and TS.

Table 11-5 shows the evolution (value of the best solution) of our *memetic* algorithm according to the number of evaluations (25,000, 50,000, 75,000 and 100,000): Note that it demonstrates its capacity to evolve and improve regarding all functions except for the function 6. Also, note from Table 11-4 and 11-5 that with 50,000 evaluations (and even with 25,000) our *memetic* performs better than other strategies (except in function 5).

Table 11-5. Evolution of the best results of ME

Function	25,000	50,000	75,000	100,000
1	1.86E-13	7.13E-14	6.45E-14	6.32E-14
2	1.67E-03	2.10E-04	1.20E-04	6.56E-05
3	2.30E-03	7.05E-04	6.20E-04	4.62E-04
4	6.02E-02	5.37E-03	4.24E-03	1.93E-03
5	3.43E+00	1.12E+00	4.92E-01	4.58E-01
6	1.51E-04	1.51E-04	1.51E-04	1.51E-04

Table 11-6 presents the mean and maximum values of the results. Even the worst results of our ME are competitive when compared to the best results of other strategies. This indicates the consistency of our method.

Table 11-6. Mean and maximum values of E (*n_evaluations* = 100,000)

Function		ME
1	Mean	9.70E-13
	Maximum	2.28E-12
2	Mean	2.15E-03
	Maximum	1.09E-02
3	Mean	1.35E-02
	Maximum	5.85E-02
4	Mean	2.17E-02
	Maximum	9.12E-02
5	Mean	3.96E+00
	Maximum	1.35E+01
6	Mean	7.84E-04
	Maximum	1.10E-03

Subsequently, a test set of 150 vectors was generated where the input values were generated with the same distribution as the training set. This test set was used to examine the capacity of the network to predict values of *y* when the values of *x* are not in the training set. Table 11-7 shows the results achieved in the test set for the weights obtained by the algorithms. The results obtained with our *memetic* algorithm for the test set are not significantly worse than for the training set.

Table 11-7. Best value of *E* obtained for BP, SA, GA, TS and SS for the test set

Function	BP	SA	GA	TS	SS	ME
1	1.76E+00	1.56E-05	3.68E-07	2.79E-06	9.37E-07	**1.13E-12**
2	9.11E+01	1.85E-01	1.72E-02	1.96E-01	2.13E-01	**1.45E-04**
3	2.34E+01	5.53+00	1.38E+00	1.69E+00	1.55E+00	**9.89E-02**
4	4.31E+02	2.73E+00	4.53E-02	1.15E+00	9.26E+00	**6.52E-03**
5	5.28E+04	1.36E+00	**2.02E-02**	5.68E-02	5.98E+03	2.43E+00
6	1.95E-01	2.69E-01	3.40E-02	6.75E-02	2.90E-01	**8.64E-03**

7. FINAL CONCLUSIONS AND CONSIDERATIONS

In this chapter we proposed a new Memetic algorithm for the training of artificial neural networks. This algorithm combines traditional genetic operators with an improvement method –a hillclimber– specially designed for this problem. In our approach we rely heavily on self-adapting strategies of the search parameters. For example, the mutation rate of the underlying genetic algorithm is adapted on-line accordingly to the state of the search. In addition, the hill-climber has a variable radius search strategy that either diversifies or intensifies the search accordingly to which region of the search space the algorithm is currently sampling. Both these self-adapting mechanism are derivative independent and as such could be use in many other neural network topologies.

The following features of our approach are worth highlighting:

- It generates good solutions in very short time (in this case, the number of evaluations), which makes it suitable for online learning procedures.
- It outperforms or equals the results obtained with other strategies proposed in recent works.
- It shows the capacity to evolve and to not "stall" when more computational time is available.
- The results obtained in the test set do not undergo significant deterioration in relation to the training set, and it continues to be better than the other strategies in almost all cases, thus showing it generalization capabilities.

Thus, the new algorithm fulfils the aims set out in Section 2. From a more general point of view we can say that the method proposed -as well as other metaheuristic methods- are often more convenient than traditional gradient-based methods for a number of reasons: first, they are not restricted transfer functions for which gradient information is available, second, metaheuristic methods such as our memetic algorithm can also handle continuous cases, third, population based training methods are more unlikely to get stuck in poor local optimae like some gradient based methods.

ACKNOWLEDGEMENTS

The authors are grateful for financial support from the Spanish Ministry of Education and Science (National Plan of R&D - Project SEJ2005-08923/ECON).

REFERENCES

Baldwin, J., 1896, A new factor in evolution, *American Naturalist* **30**:441-451.

Dawkins, R., 1976, *The Selfish Gene*, Oxford University Press, Oxford, UK.

Dorsey, R. E., Johnson, J. D., and Mayer, W. J., 1994, A genetic algorithm for the training of feedforward neural networks, in: *Advances in Artificial Intelligence in Economics, Finance and Management*, vol. 1, J. D. Johnson, A. B. Whinston, eds., JAI Press, Greenwich, CT, pp. 93-111.

Eshelman, L., 1990, The CHC adaptive search algorithm: how to have safe search when engaging in non-traditional genetic recombination, in: *Foundations of Genetic Algorithms*, G. Rawlins, ed., Morgan Kaufmann, San Francisco, pp. 263-283.

Freisleben, B., and Merz, P., 1996, A genetic local search algorithm for solving the symmetric and asymmetric travelling salesman problem, in: *Proceedings of the IEEE Conference on Evolutionary Computation*, IEEE Press, Piscataway, NJ, pp. 616-621.

Gallant, R. A., and White, H., 1992, On learning the derivatives of an unknown mapping with multilayer feedforward networks, in: *Artificial Neural Networks*, Blackwell, Cambridge, MA, pp. 206-223.

Glover, F., 1989, Tabu search, *ORSA Journal on Computing* **1**:190-206.

Goffe, W. L., Ferrier, G. D., and Rogers, J., 1994, Global optimization of statistical functions with simulated annealing, *Journal of Econometrics* **60**:65-99.

Hart, W., 1994, *Adaptive Global Optimization with Local Search*, PhD thesis, University of California, San Diego, USA.

Hart, W. E., Krasnogor, N., and Smith, J. E., 2004, *Recent Advances in Memetic Algorithms*, Series Studies in Fuzziness and Soft Computing, Springer–Verlag.

Hart, W., De Laurentis, J., and Ferguson, L., 2003, On the convergence of an implicitly self-adaptive evolutionary algorithm on one-dimensional unimodal problems, *IEEE Trans. on Evolutionary Computation* (to appear).

Hinton, G., and Nowland, S., 1987, How learning can guide evolution, *Complex Systems* **1**:495-502.

Houck, C., Joines, J., Kay, M., and Wilson, J., 1997, Empirical investigation of the benefits of partial Lamarckianism, *Evolutionary Computation* **5**:31-60.

Ichimura, T., and Kuriyama, Y., 1998, Learning of neural networks with parallel hybrid GA using a royal road function, in: *IEEE International Joint Conference on Neural Networks*, IEEE, New York, NY, vol. 2, pp. 1131-1136.

Kallel, L., Naudts, B., and Reeves, C., 2001, Properties of fitness functions and search landscapes, in: *Theoretical Aspects of Evolutionary Computing*, L. Kallel, B. Naudts and A. Rogers, eds., Springer, Berlin, Heidelberg, New York..

Kirkpatrick, S., Gelatt, C., and Vecchi, M., 1983, Optimization by simulated annealing, *Science* **220**:671-680.

Knowles, J., and Corne, D., 2001, A comparative assessment of memetic, evolutionary and constructive algorithms for the multi-objective d-mst problem, in: *2001 Genetic and Evolutionary Computation Workshop Proceeding*, pp. 162-167.

Krasnogor, N., 2002, *Studies in the Theory and Design Space of Memetic Algorithms*, PhD thesis, University of the West of England, Bristol, U.K.

Krasnogor, N., 2004, Self-generating metaheuristics in bioinformatics: The protein structure comparison case, in: *Genetic Programming and Evolvable Machines*, Kluwer academic Publishers, vol. 5, pp. 181-201.

Krasnogor, N., Blackbourne, B., Burke, E., and Hirst, J., 2002, Multimeme algorithms for protein structure prediction, in: *Proceedings of the Parallel Problem Solving From Nature*, Lecture Notes in Computer Science, Springer, pp. 769-778.

Krasnogor, N., and Gustafson, S., 2004, A study on the use of ``self-generation'' in memetic algorithms, *Natural Computing* **3**:53-76.

Krasnogor, N., and Pelta, D., 2002, Fuzzy memes in multimeme algorithms: a fuzzy-evolutionary hybrid, in: *Fuzzy Sets based Heuristics for Opt.*, J. Verdegay, ed., Springer.

Krasnogor, N., and Smith, J., 2000, A memetic algorithm with self-adaptive local search: TSP as a case study, in: *Proceedings of the Genetic and Evolutionary Computation Conference*, D. Whitley, D. Goldberg, E. Cantú-Paz, L. Spector, I. Parmee and H. G. Beyer, eds., Morgan Kaufmann, San Francisco, pp. 987-994.

Krasnogor, N., and Smith, J., 2001, Emergence of profitable search strategies based on a simple inheritance mechanism, in: *Proceedings of the Genetic and Evolutionary Computation Conference*, L. Spector, E. Goodman, A. Wu, W. Langdon, H. M.Voigt, M. Gen, S. Sen, M. Dorigo, S. Pezeshk, M. Garzon, and E. Burke, eds., Morgan Kaufmann, San Francisco, pp. 432–439.

Krasnogor, N., and Smith, J., 2005, A tutorial for competent memetic algorithms: Model, taxonomy and design issues, *IEEE Trans. on Evolutionary Computation* **9**(5):474-488.

Laguna, M., and Martí, R., 2002, Neural network prediction in a system for optimizing simulations, *IEE Transactions on Operations Engineering* **34**(3):273-282.

Land, M., 1998, *Evolutionary Algorithms with Local Search for Combinatorial Optimization*, PhD thesis, University of California, San Diego, USA.

Lecun, Y., 1986, Learning process in an asymmetric threshold network, in: *Disordered Systems and Biological Organization*, Springer, Berlin, pp. 233-240.

Mayaley, G., 1996, Landscapes, learning costs and genetic assimilation, *Evolutionary Computation* **4**(3):213-234.

Merz, P., 2000, *Memetic Algorithms for Combinatorial Optimization Problems: Fitness Landscapes and Efective Search Strategies*, PhD thesis, Department of Electrical Engineering and Computer Science, University of Siegen, Germany.

Merz, P., and Freisleben, B., 1999, Fitness landscapes and memetic algorithm design, in: *New Ideas in Optimization*, D. Corne, M. Dorigo and F. Glover, eds., McGraw Hill, London.

Montana, D. J., and Davis, L., 1989, Training feedforward neural networks using genetic algorithms, in: *Proceedings of the Third International Conference on Genetic Algorithms*, Morgan Kaufmann, San Mateo, CA, pp. 379-384.

Moscato, P., 1989, On evolution, search, optimization, genetic algorithms and martial arts: Towards memetic algorithms, Caltech Concurrent Computation Program, C3P Report 826.

Moscato, P., 1993, An Introduction to Population Approaches for Optimization and Hierarchical Objective Functions: A Discussion on the Role of Tabu Search, *Annals of Operations Research*, **41**(1-4):85-121.

Moscato, P., 2001, *Problemas de Otimizaço NP, Aproximabilidade e Computaçao Evolutiva: Da Pratica a Teoria*, PhD thesis, Universidade Estadual de Campinas, Brasil.

Parker, D., 1985, Learning logic, Technical Report TR-87, Center for Computational Research in Economics and Management Science, MIT, Cambridge. MA.

Rudolph, G., 1996, Convergence of evolutionary algorithms in general search spaces, in: *Proceedings of the International Congress of Evolutionary Computation*, pp 50-54.

Rumelhart, D., Hinton, G., and Williams, R., 1986, Learning internal representations by error propagation, in: *Parallel Distributed Processing: Explorations in the Microstructure of Cognition*. Vol.1. D. Rumelhart and J. McCleeland, eds., MIT Press. Cambridge. MA.

Schaffer, J. D., 1994, Combinations of genetic algorithms with neural networks or fuzzy systems, in: *Computational Intelligence: Imitating Life*, J. M. Zurada, R. J. Marks and C. J. Robinson, eds., IEEE Press, pp. 371-382.

Schaffer, J. D., Whitley, D., and Eshelman, L. J., 1992, Combinations of genetic algorithms and neural networks: A survey of the state of the art, in: *COGANN-92 Combinations of Genetic Algorithms and Neural Networks*, IEEE Computer Society Press, pp. 1-37.

Sexton, R. S., Alidaee, B., Dorsey, R. E., and Johnson, J. D., 1998, Global optimization for artificial neural networks: A tabu search application, *European Journal of Operational Research* **106**:570-584.

Sexton, R. S., Dorsey, R. E., and Johnson, J. D., 1999, Optimization of neural networks: A comparative analysis of the genetic algorithm and simulated annealing, *European Journal of Operational Research* **114**:589-601.

Topchy, A. P., Lebedko, O. A., and Miagkikh, V. V., 1996, Fast learning in multilayered networks by means of hybrid evolutionary and gradient algorithms, in: *Proceedings of International Conference on Evolutionary Computation and its Applications*, pp. 390-398.

Turney, P., 1996, How to shift bias: lessons from the Baldwin effect, *Evolutionary Computation* **4**(3):271-295.

Vavak, F., Fogarty, T., and Jukes, K., 1996, A genetic algorithm with variable range of local search for tracking changing environments, in: *Proceedings of the 4th Conference on Parallel Problem Solving from Nature*, Lecture Notes in Computer Science 1141, H. M. Voigt, W. Ebeling, I. Rechenberg and H. P. Schwefel, eds., Springer, pp. 376-385.

Werbos, P., 1974, *Beyond Regression: New Tools for Prediction and Analisys in the Behavioral Sciences*, PhD thesis,. Harvard, Cambridge.

Whitley, L., Gordon, S., and Mathias, K., 1994, Lamarkian evolution, the Baldwin effect, and function optimization, in: *Proceedings of the 3rd Conference on Parallel Problem Solving from Nature*, Lecture Notes in Computer Science, vol. 866, Y. Davidor, H. P. Schwefel, and R. Manner, eds., Springer, pp. 6-15.

Wolpert, D., and Macready, W., 1997, No free lunch theorems for optimization, *IEEE Transactions on Evolutionary Computation* **1**(1):67–82.

Yao, X., 1993, Evolutionary artificial neural networks, *Int. Journal of Neural Systems* **4**(3):203-222.

INDEX

N

O

P

Q

R

S

T

U

V

W